# About Island Press

Since 1984, the nonprofit organization Island Press has been stimulating, shaping, and communicating ideas that are essential for solving environmental problems worldwide. With more than 1,000 titles in print and some 30 new releases each year, we are the nation's leading publisher on environmental issues. We identify innovative thinkers and emerging trends in the environmental field. We work with world-renowned experts and authors to develop cross-disciplinary solutions to environmental challenges.

Island Press designs and executes educational campaigns, in conjunction with our authors, to communicate their critical messages in print, in person, and online using the latest technologies, innovative programs, and the media. Our goal is to reach targeted audiences—scientists, policy makers, environmental advocates, urban planners, the media, and concerned citizens—with information that can be used to create the framework for long-term ecological health and human well-being.

Island Press gratefully acknowledges major support from The Bobolink Foundation, Caldera Foundation, The Curtis and Edith Munson Foundation, The Forrest C. and Frances H. Lattner Foundation, The JPB Foundation, The Kresge Foundation, The Summit Charitable Foundation, Inc., and many other generous organizations and individuals.

The opinions expressed in this book are those of the author(s) and do not necessarily reflect the views of our supporters.

# Thicker Than Water

Black-footed albatross spotted over the eastern North Pacific Gyre while aboard *S/Y Christianshavn* in 2016. Photo by Erica Cirino.

# Thicker Than Water

## THE QUEST FOR SOLUTIONS
## TO THE PLASTIC CRISIS

Erica Cirino

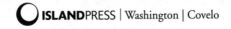

ISLANDPRESS | Washington | Covelo

Library of Congress Control Number: 2021935869

All Island Press books are printed on environmentally responsible materials.

Manufactured in the United States of America
10  9  8  7  6  5  4  3  2  1

*Keywords:* albatross, Bakelite, bioplastic, Cancer Alley, eastern North Pacific Gyre, Great Pacific Garbage Patch, gyres, Mariana Trench, microplastic, Midway Atoll, nanoplastic, petrochemicals, plastic ban, Plastic Change, PFAS, plastic industry, plastic pollution, recycling

# Contents

# Foreword

by Carl Safina

I recently got an appeal from the environmental group Oceana asking, "When did you first become aware of the ocean plastic pollution crisis? For many people, it all started in 2015 with a viral video of a sea turtle with a plastic straw lodged in his nostril."

Well, hmm. Let me tell you how, for me, "it all started" decades ago.

When I was a kid living in a city apartment, my family got deliveries of milk and soda. We'd put our empties in the crate in the hallway, and the delivery person would leave the new bottles—and take the empties. The empties weren't recycled. Nor were the "new" bottles all new. Empty bottles got sterilized and reused. The new milk and soda came in bottles that were sometimes brand new but usually scratched from repeated use.

Many things that now seem hard to imagine in anything but plastic came in, or were made from, other materials. Glass. Metal. Waxed paper and waxed cardboard.

Waste was considered unethical. People wanted *not* to waste things, and it was cheaper not to. My parents had gone through the Great

Depression, and the scarcities they endured made them appreciate what they had. My father would sometimes say, "Waste not, want not." I looked up the origin of that phrase, and here's what popped up:

> This adage was quoted—and perhaps coined—by Maria Edgeworth (*The Parent's Assistant*, 1800) . . . It was widely repeated throughout the nineteenth century, but has been heard less in the current throwaway society.[1]

The current throwaway society, indeed. Getting from a culture of reusables to a throwaway society didn't happen by accident. Because plastic was wasteful, advertising campaigns had to accustom consumers to the idea. It took a very concerted effort, over years. I remember when companies started advertising wastefulness as a virtue and something to be desired. The ads made a massive new push for a plastic revolution. "Use once; throw away" was one common tagline. "Disposable" was another.

A few things were better in plastic. It was no fun accidentally breaking a glass shampoo bottle in the shower, and TV ads energetically demonstrated plastic as safer.

Plastic soon began replacing all kinds of things. Success of the revolution was such a sure bet that in the classic movie *The Graduate* (1967), the twenty-one-year-old Benjamin Braddock (Dustin Hoffman) receives this now-infamous advice about the direction his future should take: "One word: plastics."

Back in the real world, we watched plastic overtake our material existence. Soon you could get yogurt, a snack with a two-week shelf life, packaged in an eternal material that could break up physically but never break down chemically. But I don't recall anyone suggesting that throwing plastic away was a problem. Until—

Just a few years later, in the 1970s, birds, turtles, and other wildlife started to turn up tangled in plastic six-pack rings. The conscientious among us learned to snip the rings before throwing them away. But there were two problems. The *un*conscientious weren't snipping, and they weren't careful about throwing things away.

By the 1990s, it was clear to some that plastics were building up in the ocean and washing ashore on the most distant coasts. By the early 2000s, on Laysan Island and Midway Atoll, as far from the continents as it is possible to be, I saw dead albatrosses full of plastic. I'd seen an albatross try to feed a toothbrush to her chick. By then, everywhere I went, no matter how distant from people, even wilderness shores of Alaska, I was continually amazed and dismayed at the plastic building up. Whales and turtles were becoming tangled in plastic, or dying after eating it.

Glaring as the problem seemed to those of us who were experiencing it, it remained out of sight and mind to nearly everyone else. When Captain Charles Moore invited me to go on a trip to witness firsthand the North Pacific Gyre, which he'd dubbed the Great Pacific Garbage Patch, I tried to interest *National Geographic* in the story. They declined, saying—of all things—that plastic in the ocean "wasn't visual enough." Can you imagine? A few years later, the tragedy of ocean plastic was, very visually, *National Geographic*'s cover story. It has become so inescapable, it's even in our seafood.

Though I haven't sailed to the North Pacific Gyre, author Erica Cirino has. Picking up the story and making it her business, she's traveled to many seas and shores to bear witness. Here, in *Thicker Than Water*, she takes us with her to distant oceans and faraway coasts. Erica Cirino shows what it's like to live on the high seas while sleuthing into a quiet, monumental problem. Up close and in person, Erica lets us see, feel, taste, and smell what the ocean plastic crisis is, what it means. She shows why recycling has failed, and how oil and gas companies (which sell the fossil fuels that plastic is made from) have helped make sure recycling continues to fail. Erica brings us up to date on a problem that continues to mount but—as she shows us—is not insurmountable.

Plastic is a problem we make. It's a problem we'll solve. Already, municipalities and countries are passing laws to limit the use of throwaway plastics. And let's not forget the most important fact: Plastic is made by people. It doesn't have to be. Plastic has been in commercial production for only about eighty years. New companies are developing new materials that can give you your yogurt and perform like plastic in every way except that they don't last forever. The answers lie in ending planned wastefulness and developing new materials for a post-carbon, post-plastic world.

With feeling and with flare, in *Thicker Than Water* Erica Cirino takes us into the problem—and shows the way out.

# Preface
# Out to Sea

The island I grew up on doesn't feel like an island. On the west end is the hipster mecca of Brooklyn, in all its vinyl-listening-Mason-jar-drinking-flannel-wearing glory. On the east end is the rich-and-famous playground of the Hamptons—trust funded, gold encrusted, and fashionable.

I've always lived somewhere in the middle of these two places, always just a short walk from saltwater—and sure, growing up I spent a lot of time at the beach, plucking shells and glass treasures from the shore, splashing in the waves, kayaking and racing around the harbor in my Laser sailboat. But it wasn't until I started offshore sailing, when I was twenty-four years old, that the ocean really seeped into my identity. To date, I've crossed at least ten thousand nautical miles by sea, into the North and South Pacific, across the Atlantic, around the northwestern coast of Iceland, and on other journeys.

And in getting to know the sea, I've recognized her critical importance as a selfless provider. She absorbs carbon dioxide and releases oxygen so we can breathe. She contains hundreds of thousands of plant and animal species, many of which are sources of food and medicine that we rely on to survive. She provides us with a way to ship the things we buy and use, from toothbrushes to cars. She safeguards minerals and oil that have historically powered modern human society. She captivates our imaginations, sparks our creativity, and is a core part of our culture. The sea is a place we go to play and pray.

As the ocean gives to us, we take from her with abandon. We've taken more than our share of oxygen, of plants and animals, of minerals and oil. And when we have given to the sea, it's been all the wrong things: More carbon than she can cope with, causing acidification and its consequent massacre of coral reefs and any species with a calcium carbonate shell. More boat traffic than she can handle, leading to marine mammals' deadly and disfiguring collisions with ship propellers. More acoustic military drills and bombings than her resident marine wildlife can bear, causing behavioral anomalies in whales, fish, and dolphins, which rely on sound to survive. More oil spills and nuclear meltdowns than she can easily shake off. And more plastic debris than she has room to hold; what eighty years ago was an unknown phenomenon today has turned into one of the worst environmental crises in history.

While plastic is a material made on land, my story about humanity's plastic crisis begins in the Pacific Ocean's notorious Great Pacific Garbage Patch, where so much of our detritus is accumulating to the detriment of marine plants and animals. This single voyage compelled me to dedicate the last five years and counting to covering the story of our global plastic disaster, by sea and by land; documenting pollution and getting to know the many people who are working feverishly to address the crisis before it is too late—for the oceans, and, as I have learned, all of us.

Out at sea, time is not measured in hours or minutes, but by the intensity of the burning sun, the oscillating fade-sparkle-fade of thousands of stars and specks of glowing algae, the size and shape of the moon, the furor or calm of the sea. Out there, the distractions of a modern life are abandoned on land, leaving one with nothing but her soul and most vivid dreams—and most tormenting demons.

Out there, I learned, life is beautiful and wild and painful, and in its pure rawness, the sea has the potential to reveal the truth. The sea can show us what it is in life we need, and what we can live without.

Kristian Syberg, professor and noted plastic pollution researcher at Roskilde University in Denmark, searches for plastic and other human-made debris to scoop from the surface of the eastern North Pacific Gyre while aboard *S/Y Christianshavn* in 2016. Photo by Erica Cirino.

PART I

# The Missing Plastic

CHAPTER 1

# Welcome to the Gyre

Big changes happen fast here on the gyre's edge. Looking out over the wild, whipping expanse of sea before me, I gripped the wheel of the fifty-four-foot steel sloop and braced myself. In this half-dark, half-dazzling sunrise hour, the seas had transformed from simply precarious to volatile and violent. The gale had moderated over the past few days since our departure from Los Angeles, when it had been blowing hard. But now it kept changing direction, forcing me to pay close attention to the red plastic arrow spinning wildly atop the mast, and the telltales on the mainsail, as I worked to press in as close to the wind as possible. When I fell into a good rhythm, the mainsail and jib were tight and protruding, their bellies puffed full of wind; we were moving fast over the churning sea.

Between the craggy curve of the Californian coast and Hawai'i's long chain of volcanic islands exists a clockwise-spinning vortex of seawater known as the eastern North Pacific Gyre. When you're sailing into it, you can't see the water turning, but you can feel the elements of the sea coming together to create the turbulence that fuels it. It may seem like no sea captain of a sound mind would choose to sail through the gyre, with turbulent waters at its edge, and a near-windless no-man's-land inside. Yet that was exactly where we planned to sail, led by captain Torsten Geertz and the ship's co-owner, a one-man tornado of energy named Henrik Beha Pederson.

The more I learned about plastic pollution, the more I felt the need to see this infamous Garbage Patch for myself. It is actually one of two distinct garbage patches accumulating on either side of the North Pacific Gyre; another area of highly concentrated trash spins, smaller, farther west, off the coast of Japan. Much trash is carried between the two patches, over a colossal area of ocean.[1] I boarded the sailboat and was soon facing into the gyre with the rest of the crew.

As the early morning wore on and we ventured farther into the gyre, the contrasting black-and-orange dawn sky grew more orange and less black, and then morphed to yellow to pink to purple, flipping through the pages of a Pantone color book until it settled on a uniform cerulean shade. At the same time, the sun crept up from behind the horizon until it was suspended in the sky, and the choppiness of the dawn sea subsided. I exhaled and relaxed my grip on the wheel.

Quiet. There was so much quiet out at sea. Any noises that were present were rhythmic, natural, easy to acclimate to—noises quickly woven into the fabric of your existence: the smooth *phsssssh-phsssssh-phsssssh* of the steel hull cutting through gentle waves; the repetitive *pa-pa-pa-pa-pa-pa-pa* of the sails flapping every time the wind died down or changed direction; the rattling *clink-clink-clink-clink-clink* of the mainsail shackles on the tall aluminum mast when a squall snuck up on our ship; the occasional *grrrrrrrunk* of the wheel turning around its central axle, which apparently needed some grease. And then, immediately around me, there were the sporadic human elements of life at sea: breath, movement, and speech.

"Do you see that?" a voice cut through the calm. It was Malene Møhl, a Copenhagen-based plastic researcher with a love of sailing. She squinted her hazel eyes, watching the waves. The seven other crew members were either still asleep in their slim wooden bunks or milling about inside the ship's cramped living quarters while Malene and I carried out our overlapping early morning shifts on deck, minding the sails and navigation.

"Look, look, off the bow!" Malene said. She motioned a quick hand toward the water. About ten meters in front of our ship was the shredded corner of a sun-bleached orange plastic fish crate, suspended in the curling blue arc of a wave. Minutes later, I saw a fist-sized chunk of white Styrofoam drift by the ship's starboard side, and then a small pink plastic dustpan off port. Next there went a punctured green plastic shampoo bottle, and then past the bow a barnacle-encrusted Tupperware lid. Soon after, it was pure blue sea again. Henrik, the crew's organizer, scrambled up the short wooden stairs from the hull to the cockpit and raced to the bow while clicking on a self-inflating life vest. His blond hair was mussed, his eyes encircled by shadows indicating a lack of sleep.

Once on the bow, Henrik snapped to life standing beneath the luffing, lazy genoa, calling out a blow-by-blow report of the items floating by. "A rope! A fish crate! A tube! A bottle! A balloon!" One by one, the rest of the crew, awakened and alerted by the sound of Henrik's booming voice, climbed up onto the deck, and they too began watching the waves.

After a slow but steady stream of plastic items would intercept the ship for a few minutes, we'd see nothing, and then a few minutes of plastic again, and then nothing, and then the pattern would repeat. When the items ventured close enough to the ship, Henrik would lean over the metal railing and scoop them up with a large fishing net. After about an hour, he had stacked a shin-high pile of colorful, barnacle-encrusted trash on the deck. And that would turn out to be only a small part of the day's plastic haul. We were at least one thousand nautical miles in any given direction from landmasses inhabited by humans and their plastic societies.

Plastic was the whole reason Henrik had brought the ship—an old steel sloop called *Christianshavn*—and crew out into this desolate part of the sea. He's a biologist by training, one who has studied the effects of humanity's use of plastic on wildlife and the environment. But he's a sailor at heart. During many pleasure trips spent sailing in exotic places like Greece and Thailand with *Christianshavn*'s Danish co-owners, Henrik witnessed enormous amounts of plastic items commonly used on land floating around in the ocean and washing up on even some of the most remote shores. It was then he realized it was time to repurpose *Christianshavn* from a timeshare vacation ship into a research vessel. In late 2012, he established a nongovernmental organization called Plastic Change, focused on shifting the world's relationship to its most beloved material, something he viewed as one of the world's foremost environmental and social problems, and a problem that he as a scientist-sailor and former Greenpeace manager might be well equipped to address.

"Plastic defines our culture," Henrik declared in 2014, at an early board meeting for his nascent nonprofit. "We must not let it define our future." That year, he commandeered *Christianshavn* to carry out research in the oceans, collecting data on marine plastic pollution by scooping it out of seas and trying to answer questions about each piece—like where in the world it came from, what it had been used for, and how much other plastic was out *there*, in the oceans. Henrik hoped sharing his organization's at-sea findings would compel others to care, and ultimately, take action—though at the time, it was less clear what appropriate action should look like.[2]

By the time I boarded *Christianshavn* in Los Angeles in November 2016, Henrik had directed the ship's scientific voyage from his home waters outside Denmark through the Mediterranean, across the Atlantic into the Caribbean, then through the Panama Canal to Colombia, around the Galápagos Islands, to Mexico, up to Los Angeles, and—to kick off its grand finale in the Pacific—into the most notoriously plastic polluted stretch of ocean in the world: the eastern North Pacific Gyre.[3]

And so, there we were, in a part of the ocean so polluted it's been nicknamed the Great Pacific Garbage Patch. Yet it became rapidly apparent while gazing out over these waters that "the patch" was not really a static, floating pile of plastic, as the so-often sensationalistic global media machine has commonly portrayed it. The reality is much graver: These waters are more akin to a soup to which humanity has added an unknown number of plastic items and pieces.[4] The plastic is commonly suspended right below the surface, pushed just out of sight, constantly and unpredictably stirred by the roiling sea.

When it seemed that the bulk of the morning's plastic parade had dispersed, so did the sailors around the ship. As the sky warmed and brightened from dawn to day, one by one the crew gathered in the snug cockpit and settled on the double-tiered teak benches, shoulder to shoulder. They sat barefoot, wearing mismatched athletic shorts, T-shirts and waterproof jackets, topped with slim self-inflating life vests. Someone carried up a hot thermos of coffee; two loaves of slightly burnt, misshapen whole-grain bread, baked the day before in the ship's tiny and greasy oven; a jar of store-bought strawberry preserves; and a tub of butter, quickly liquefying in the subtropical warmth. The hot black coffee tasted bitter and slightly marine when sipped from our saltwater-washed ceramic mugs; the bread—made from dough mixed with one-quarter saltwater in an attempt to conserve our freshwater supply—was crunchy, and similarly salty, infused with the essence of the sea.

It was Henrik's turn at the wheel. Each day we were required to work two four-hour sailing shifts, set eight hours apart. Each sailor's sched-ule—when she had time to sleep and eat and socialize—was largely dictated by the timing of the day's sailing duties. After a 2:00 a.m. to

6:00 a.m. shift, for example, I could muster little more than to slide into my wooden, sarcophagus-shaped bunk.

That day, however, everyone was awake and sitting in the cockpit by late morning. As we sat together, Malene noticed something round and big and green and submerged ebbing toward our ship. She followed it with her eyes and an unwavering finger as it bobbed in the waves. Henrik's hands flitted across the spokes of *Christianshavn*'s blue wheel as he nosed the ship in the direction of whatever was in the water. The sailors scrambled onto the bow watching and wondering—was it a swimming sea turtle, fish, or—? No, the deadened manner in which it moved indicated this thing, whatever it was, was not a living creature.

The sailors' faces shifted from expressions of curiosity to disgust as they greeted a massive tangle of green, orange, black, and white ropes and fishing nets. The captain, Torsten Geertz, cut through the huddled people peering over the side of the ship and swiftly swung a gaff down into the waves. Malene helped him wrestle the unwieldy tangle of plastic fishing gear—a common phenomenon that emerges when plastic ropes and nets meet others in the oceans—onto the deck. She stood over it, squinting in the intensifying late-morning light, counting dozens and dozens of separate pieces of gear.

Tangles like this are often referred to as "ghost" fishing gear for their ability to catch and kill marine wildlife long after having been decommissioned, lost, or inappropriately discarded at sea.[5] It appeared this ghost gear had not recently entrapped anyone.

Malene told us that this wasn't the first time Plastic Change had encountered ghost gear. A few months earlier, she had been aboard *Christianshavn* as the crew sailed from Mexico to Los Angeles, cruising at a good clip when they'd spotted some dolphins splashing around. Then they noticed a seabird called a booby, which seemed to be standing on the water right above something colorful. To solve the mystery of the bird that could walk on water, the crew changed course and made a slow approach.

"We learned that it was a fishing net, jumbled with lines, ropes, and empty plastic jugs used for floatation," said Malene. "And in that mess was a scruffy-looking sea turtle. Then we realized: the dolphins and the booby were hanging around to feed on the fish and crabs that had gathered underneath this shelter, this floating island—this entangled turtle—like they would do if they'd found a seaweed raft."

On first glance, the sailors took the turtle for dead. A garden of slick seaweed and puffs of algae grew off the caramel-colored scutes of his shell. He had probably been trapped, possibly unable to eat, and certainly prohibited from moving by his own free will—instead he had been carried by water currents for quite some time. But then they noticed the turtle lift his head a tiny, nearly imperceptible, amount.

The sailors emptied *Christianshavn*'s full sails, slowing the vessel to a stop alongside the sea turtle. One sailor jabbed a gaff into the clump of barnacle-encrusted ropes and nets that ensnared the turtle's shell, while another leaned his body out over the side of the boat and painstakingly cut the trapped creature free. They pulled the sliced-up mess of plastic bottles, net, and rope on deck as proof of the rescue, to warn others of the serious threat plastic poses.

"As soon as the last rope was severed, the sea turtle swam off a ways before looking back 'over his shoulder' and sent a flipper up as if to say goodbye," Malene said. "We presume he got a second chance to live. Most animals who swim into ghost nets are not as lucky."[6]

Each year, half a million—or maybe more—marine animals become entangled in ghost gear, their plastic-bound cadavers observed at sea and washed up on beaches. Many more die unseen. Those who cannot escape are forced to carry massive tangles of gear on their bodies, if they are able to move at all. Sometimes entangled marine animals are able to shake their gear over time, but the scars last a lifetime. Entangled marine animals often die in a slow, painful manner as plastic lines cut into flesh, sever limbs, and restrict movement. Some creatures that must surface to breathe, like whales and sea turtles, or who dwell mainly above water, like seabirds, drown when ghost gear weighs them down beneath the waves. Few creatures escape ghost gear physically unscathed, and scientists are just beginning to reflect upon the psychological traumas almost certainly inflicted on individuals who are or who have been entangled.[7]

In the past, much fishing gear was smaller and made from cotton and other natural fibers that decomposed rapidly in the oceans.[8] Modern fishers' predominantly plastic nets and lines—when employed industrially—can run as long as thirty miles in some parts of the oceans, creating an often-inescapable hazard to creatures in their path. And fisheries scientists have evidence that fishing fleets around the world lose or intentionally discard roughly 6 percent of their nets, 9 percent of traps, and 30 percent of lines annually, adding perhaps a million tons of ghost gear—like that we encountered aboard *Christianshavn*—to the sea each year.[9]

We had confiscated a deadly weapon from the ocean. It was the least we could do. Malene and Torsten dragged the heavy ghost gear to a nook on deck out of the way of our foot traffic. After a few days, we noticed the net began to reek. Upon closer inspection, we could see that dozens of tiny brown crabs were hitchhiking inside the knot of algae-coated ropes. Malene carefully examined the net for any surviving stowaways and flicked as many as she could find overboard.

As he had done with the ghost gear pulled from the entangled turtle in waters off the coast of Mexico, Henrik planned to show this ghost net to the public once we got to shore. What good was bearing witness to so much destruction if we did not share our findings with those on land who may never have a chance to go out to sea, to see for themselves?

The oceans are rife with evidence that plastic exists at odds with natural life, like so many human inventions. Plastic can be deadly. It's also become one of the most ubiquitous materials on Earth, used to hold all manner of food and drink we consume; to make everything from lifesaving medical devices to fibers woven into the clothing that hangs on our backs. Virtually all plastic made today is derived from petrochemicals, like ethanol and phenol, pulled from natural gas and crude oil.[10] In the long story of human history, tapping gas and oil on a mass scale—to make plastic but also provide other materials, chemicals, electricity, and mass transportation to modern human society—is a fairly recent development.[11] Before collective human memory of the past fades further from present, it's worth mentioning that our widespread exploitation of fossil fuels and the emergence of plastic is a story linked to a slightly older narrative about our exploitation of another substance, one only found out at sea: whale oil.

When it became apparent in the mid-1800s that the commercial whaling industry was essentially slaughtering itself out of business, people began to diminish, though not entirely sever, their reliance on whales and explored other fuel sources, including lard from pigs and tallow from cows and sheep. Camphine, a mixture of alcohol and plant oil, was another alternative, as was kerosene, a then-newly discovered fuel first derived from coal and later from distilled crude oil. Kerosene burned brightly like whale oil, without the need for whales. Its popularity among the masses drove entrepreneurs to probe the planet for fossil fuels, which they often found gushing in seeming abundance. Kerosene lit the way in humanity's massive shift in killing live plants and animals for fuel, to tapping Earth's abundant supply of energy-rich, carbon-filled fossils.[12]

Meanwhile, the consuming classes of human society confronted another mismatch between natural supply and their unnatural demand: The world's elephants were marching down the short road to extinction as the result of years of slaughter. At the time, elephant ivory was used to manufacture all manner of marketable products from billiard balls to false teeth to combs to piano keys.[13] As the price of ivory—and other then-commonly used parts pulled from the bodies of other wild animals, like tortoiseshell—increased, frustrated business owners prodded scientists to search for cheaper alternative materials.[14] In 1862, British inventor Alexander Parkes presented the world with the first manmade plastic, Parkesine, at that year's International Exhibition in London. Concocted from a mix of plant-based ingredients—including highly flammable nitrocellulose (also known as "gun cotton," a key additive in smokeless gunpowder) and solvents like alcohol, Parkesine was advertised at the exhibition as "a substance hard as horn, but as flexible as leather, capable of being cast or stamped, painted, dyed or carved." While Parkes won an award for his new material, which he molded into many luxurious products like combs and decorative bowls, Parkesine never went on to become a huge commercial success. That's probably because the material was challenging to produce consistently and was liable to explode if exposed to flame or friction. Within two years of setting out to manufacture Parkesine on a large scale, Parkes was forced to file for bankruptcy.[15]

Soon after, American inventor John Wesley Hyatt combined camphor tree oil with nitrocellulose to create another plant-based plastic, called celluloid, that could also be shaped and hardened to mimic items usually made from ivory and other animal parts.[16] Again, humans believed, as they once had when they replaced plant and animal fuels with natural gas and petroleum, that they'd outsmarted nature, this time by creating plant-based plastics—ivory, without the need for elephants. But celluloid, while successfully mass-produced to make all manner of consumer items from photographic film to table-tennis balls, was not quite the miracle consumer material industrialists had hoped for: It could be tricky to mold and tended to lose its shape when heated. Plus, like Parkesine, celluloid proved to be extremely flammable. There had to be something better out there.[17]

It was in 1907 that the significant plastic breakthrough industrialists had been wishing for finally occurred: Leo Baekeland, a Belgian-born chemist, created a plastic that completely *defied* nature because it was made not from the bodies of recently living plants or animals, but from beings that had been dead for tens to hundreds of millions of years. Working in his laboratory-garage at his "Snug Rock" home in Yonkers, New York, Leo Baekeland created the first batch of petrochemical-based plastic by exposing phenol, a component of crude oil, and formaldehyde to extreme heat and pressure in a homemade, steam-powered ovoid iron oven with a mixing arm inside—an invention he'd dub the Bakelizer. The result was a novel substance not found in nature, but instead well suited to mass production in factories—synthetic plastic he called Bakelite.[18]

"I consider this days' very successful work which has put me on the knot of several new and interesting products which may have a wide application as plastics and varnishes," Baekeland wrote in his diary on the day of his discovery.[19] Bakelite is a thermoset plastic, meaning once its chemical ingredients are mixed and hardened through curing, usually with heat, it will remain rigid.

From there on, Bakelite, marketed as "the material of a thousand uses," was molded into the forms of a countless array of affordable and durable consumer products, from ashtrays to toothbrushes, plates, toys, jewelry, napkin rings, firearms, coffins, and cars. Bakelite is distinctive: rock-hard, chunky, opaque, and often colorful. Manufacturers soon began adding fillers like asbestos, cotton, wood, and carbon black, a substance derived from the processing of fossil fuels, to Bakelite's other petrochemical ingredients in attempts to increase its strength and durability.[20]

A great variety of mass-produced, petroleum-based plastics arose following the rise of Bakelite, particularly after the end of World War II, when factories that had been churning out tanks and explosives began making plastic products. A burgeoning advertising industry promised consumers that inexpensive plastic products would make their lives more convenient, luxurious, and generally better. While padding the pockets of the rapidly growing petrochemical industry, this incredible new synthetic substance of the future—plastic—cheapened material- ism and enabled the mindset of consumption to course through human society.[21]

Those who made and sold plastic failed to point out that their material's most desirable, and thus marketable, characteristic—durability—also came with a dark downside: Because it is synthetic, made by humans in factories instead of found in nature, plastic cannot benignly decompose over time like wood, clay, rock, metal, ore, and other natural elements and substances humans have harnessed over the course of our history. At sea, wind, waves, and heat rapidly churn items made from natural materials into their various chemical components, dispersing molecules of utility throughout the sea—mainly carbon, oxygen, and hydrogen. Petrochemical-based plastic, on the other hand, does not break down into chemical elements that can be recycled. And so, plastic chokes the ocean.[22]

Plastic is so permanent because of its structure at a molecular level. All substances existing on Earth—natural and manmade, living and nonliving—are made of chemical molecules held together with elec- tricity. Think of all substances as clumps of stuck-together Lego bricks: The Lego bricks (molecules) that make up a substance snap together to create something, but with enough force, the bricks can be pulled apart, and the appearance—and sometimes chemical composition or state— of the substance will change when their bonds are broken.

We think of the plastic items we use every day—a list that, if you have kids, may include those beloved Danish plastic toy bricks, in addition to things like plastic straws and smartphones—as relatively unchanging things. But the reality is that they are made up of elemental molecules that, with enough heat, electrical or chemical energy, or physical force, *should* be capable of being pulled apart.

Yet, instead of breaking down into simple molecular components, like organic substances do, plastic breaks up into smaller and smaller pieces, pieces that remain plastic forever—as far as scientists can tell today. As soon as a plastic item is manufactured, it begins breaking up into bits. In the oceans, it may take a plastic item anywhere from a few dozen to hundreds of years, depending on the item and the conditions to which it is subjected, to completely break up into a collection of plastic particles.[23]

Plastic's inevitable breakup at sea is greatly accelerated by heat and the sun's rays, as well as the physical pounding of strong rains, waves, and wind, which break the bonds that hold plastic molecules together. Buoyant plastic floating on or just beneath the sea surface is exposed to the greatest amounts of sunlight, heat, and physical forces, so it tends to break up faster than plastic items that sink to the seafloor, where there is little to no sunlight and less movement.[24]

Plastic, the poster-child material of industrialization, was created to defy nature, to game the ephemerality of life. And so, plastic persists.

## CHAPTER 2
# Below the Surface

Across the Pacific, the bucking sloop heaved through monstrous blue undulations, sticking only vaguely to her rhumb line. Into her sails, the wind sung her instructions—no, at first she screamed her message: Move south, move south! But Torsten turned *Christianshavn*'s rudder north, and in a few moments the hulking ship was slowed to a near standstill, sails luffing in exhaustion or perhaps defiance—no, I will go *this* way!

Abruptly, we dipped into the doldrums, a near-windless area in which we'd remain for a significant portion of our journey across the eastern North Pacific Gyre. A calm, shining, blue oasis unfolded well past the horizon. There were no other ships, no other people in sight. Each wave glinted diamonds, a trick of the late afternoon light. The beauty of this place seemed ancient, everlasting. Like it never has been, and never will be, spoiled—though I was beginning to realize this perception of the ocean was quite far from the truth.

Clipped onto the ship's safety line, I crept along the narrow deck to the bow, to sit beneath the shade of the jib, one of the only places on deck shielded from the skin-searing sun. I had just a short amount of time for rest before helping to cook lunch—and later, dinner—for the whole ship. "Kitchen duty" was a responsibility that twice weekly replaced our two four-hour sailing shifts, the usual daily obligation. Rasmus Hytting, a Copenhagen-based carpenter and boat builder, was also excused from sailing duty for the rest of the day and would join me in the galley.

Inside, the ship's wooden interior was growing increasingly slick and humid, a greasy saltwater film adhering to everything and everyone inside. Rasmus and I hunted for ingredients below salt-streaked wooden floorboards, inside cupboards hidden within the saloon benches, and on the kitchen shelves—where the only security keeping their contents from cascading onto the floor were meager strips of netting. We decided on red beans and rice, one of an untold number of bean-and-rice dishes we'd consume on the journey. Besides pasta, these ingredients were among the simplest to prepare and kept well on the ship.

Rasmus lit the grime-smeared, precarious-looking gimbaled gas stove as *Christianshavn* bobbed from side to side, slowly cutting through the water. I squeezed into another part of the narrow galley, alternating my steps on the two round metal foot pumps on the kitchen floor. This started a flow of water—first saltwater and then fresh—which I directed into a Pyrex measuring cup, before adding to a dinged-up metal pot for the rice. When you set off for a long-haul sailing trip on an old and sparely equipped boat, you quickly realize every drop of freshwater at sea is extraordinarily precious. The sailors used saltwater for as many tasks as possible—cooking, bathing, and dish washing—and fresh only when brushing their teeth or to dilute the saltiness of a pot of rice or pasta and, of course, to drink.

Rasmus, who had tossed out two fishing lines every day since we'd been out at sea, would wander up to the deck every fifteen minutes or so to give them a hopeful tug and puff on a hand-rolled cigarette before returning to the galley. He was a man sparing in the words he shared with others and spent much time focused on navigating his own private thoughts. When the crew had first assembled in Los Angeles, he had introduced himself concisely: the ship's first mate, a person "focused on tight lines, full sails, and a clear deck." Despite his reserve, he and I had an easy way with each other, a mutual understanding that transcended our age—twenty-four and forty-eight—and cultural—American and Danish—differences. While I'd end up forging a special bond with everyone on the ship through our shared experience, Rasmus the reticent sailor was my first real friend in the crew.

After tackling the post-lunch cleanup, I settled on deck with a book. The sea was much calmer now that we'd passed through the edge of the gyre, and *Christianshavn* maintained a slow but steady clip. At times, life onboard was full of energy, but at other times, like this, when there wasn't much to do, the vibe was lethargic. Sofie Zander, a young woman who had studied psychology and was interested in sailing, settled down in the saloon with a dog-eared Danish novel. Sitting next to Sofie was Malene, who leafed through the latest edition of *Lonely Planet Hawaii*, researching activities to try upon our expected landfall in Honolulu. Torsten napped on the jib, which was now rolled-up on the front deck beneath the shade of the lofty mainsail and genoa. Peter Andersen, a Dane with a love of sailing then working as a cardiac researcher at Johns Hopkins University in Baltimore, napped in his bunk in the navigation room at the ship's stern. A slightly seasick Kristian Syberg, the crew's lead scientist, rested in the V-berth bunk at the bow. Rasmus sat next to me on the bench smoking cigarettes, listening to The Clash on his iPhone and watching his fishing lures dive and jump through the shining waves.

Chris Jordan, the only other American, and non-Dane, on the expedition, accompanied Henrik at the wheel. Chris and Henrik took turns steering the ship through the calm water with their heads craned to either side, on the lookout for plastic. They shouted—and Chris and I jumped up to snap photos—when plastic items floated by. Then Chris would rejoin Henrik at the wheel, while I resumed reading.

Henrik had invited Chris—an internationally renowned artist who depicts mass consumption in his artworks—to film and photograph the expedition. As an artist, Chris is perhaps best known for his 2017 movie *Albatross*, which spends much time portraying the film's namesake bird soaring across tangerine sunrise skies, dipping into cobalt blue waves, laying eggs, and raising adorably clumsy chicks, swaggering and singing to impress potential mates. He's also celebrated for a darker set of works, a series of photographs depicting dead albatrosses he's found and sliced open with a scissor on the spot—jarring scenes he also included in *Albatross*. His images, both still and moving, clearly depict the cause of the birds' deaths: guts brimming with all manner of plastic debris—lighters, bags, bottle caps, pens, forks, straws, fragments—they had plucked from the very waters we were sailing.[1]

From my perch on the bench, I noticed a cloudy white blob floating on the water's surface—hundreds of white plastic pieces, some tiny confetti-like bits and some robust, rugged chunks that look like peeled paint chips. I stuck my head over the railing for a closer look. Near the floating mass of white plastic, I noticed something else in the water—something small, finned. Something alive. I watched as a quarter-size blue-green larval fish swam to the surface, opened its tiny jaws, and swallowed a bit of white plastic the size of a pencil eraser. Within moments the fish and plastic floated away.

I felt a human presence behind me, a head peeking over my shoulder. It was Kristian. "You just got a firsthand look at how plastic gets into the food chain," he told me.

Kristian is an associate professor of environmental risk at Roskilde University in Denmark. On this expedition it was his job to help Plastic Change track down, collect, and evaluate the hazards posed by the smallest pieces of plastic that now proliferate the oceans. Environmental scientists like Kristian began to refer to these plastic particles as "microplastic" in 2004, after a team of UK-based researchers, led by Richard Thompson at the University of Plymouth, published a landmark paper coining the term to describe small bits of plastic found on beaches and in waterways.[2] By 2009, scientists investigating the plastic crisis in the oceans refined the definition of "microplastic" to include plastic particles manufactured or broken down to a diameter of five millimeters or smaller.[3]

Explaining why the eastern North Pacific Gyre and the rest of the world's oceans are filled with an infinitely increasing amount of plastic waste and its ever-shrinking plastic particles is simple: Humans manufacture and use a *lot* of plastic, an inexpensive material of convenience good at providing clothing with waterproof properties, insulating wires, and serving as a barrier against bacteria for packaged foods. It's in our electronics, toothbrushes, shoes, cigarette filters, cars, floors, paints, bags, bottles, straws, containers. Plastic is everywhere we look.[4]

While many kinds of plastic items can be reused, the world's primary use of plastic is for flimsy "disposable" items that are rarely recycled and instead are thrown away at astronomically high rates, sometimes after mere minutes of use. About 40 percent of plastic used today is actually not even really *used* by people—instead, as packaging, it covers or holds the foods and goods we purchase and is simply torn off and thrown away so we can access what's inside.[5] However, many of the plastic items we *do* use for specific purposes—like straws, cups, plates, and cutlery— are also rapidly thrown away after just one use. This quickly discarded, so-called single-use plastic is what's most likely to end up in the sea; being lightweight and flimsy, it's carried to the oceans by wind, rivers, tides, and rains when littered and dumped, especially in coastal communities. But even when plastic is handled with the best intentions—that is, deposited in waste bins, landfills, or recycling facilities—it is bound to escape, as plastic waste is swiftly accumulating and poorly contained. And so no matter where we throw away our plastic when we're done using it, much travels—often swept by elements from land into the sea.[6]

According to municipal waste records, in 1960, less than 1 percent of household trash in middle- and high-income countries was plastic, but in 2005 more than 10 percent of this household waste was plastic. Solid waste generation, which has been tightly linked to gross national income per capita, continues to rise globally.[7] And as human world population continues to skyrocket and our thirst for material and luxury goods increases, plastic's share of space in our household waste is only expected to grow into the future. And as it does so, we can expect plastic to increase its share of space in the seas.[8]

By analyzing manufacturers' data, top plastic researchers Jenna Jambeck, Kara Lavender Law, and Roland Geyer estimated that by 2015, humans had produced 8.3 billion metric tons of freshly made, non-recycled, petrochemical-based plastic since plastics were first mass-manufactured in the mid-1900s. They believe that 76 percent of that plastic was used only once or twice before being discarded.[9]

Jambeck, Law, and Geyer have noted that municipal waste data, which tends to vary from poorly kept to completely unrecorded, seems to cautiously suggest that around 79 percent of all discarded plastic has been tossed in landfills or has entered nature—air, soil, or water—after being lost or dumped there. They say it looks like about 12 percent has been incinerated, often in waste-to-energy plants. And just around 9 percent seems to have been recycled and turned into new plastic items, often combined with freshly made plastic. Plastic manufacturers are on track to having churned out thirty-four billion metric tons of new petrochemical-based plastic to fuel our consumer lifestyles by the year 2050.[10]

Crunching the numbers recorded by industries and municipalities can help scientists roughly approximate how much plastic has been created, used, and even discarded throughout history. This information may shed light on the scale of humanity's thirst for plastic. But it's another task entirely to understand our plastic addiction's full range of impacts; to do that, scientists first need to determine *where* on Earth our endless stream of mass-produced plastic has gone once it has escaped into the natural environment, particularly once it's been churned into microplastic and is no longer easily seen. The big question plastic pollution scientists are trying to answer today is, Where is all that "missing" plastic that's made its way into nature?[11]

We know there are plastic items floating in the Great Pacific Garbage Patch. We've seen them. And where these plastic items tend to collect, so do the microplastic particles they shed. But spotting individual pieces of microplastic on the open sea is difficult to do using your eyes alone. And who knows how much is lurking beneath the surface?

Like most scientists who have devoted their careers to these questions, Kristian had acquired an assortment of equipment to help him more easily detect the microplastic around him. Plastic pollution research is a still-developing area of science that's emerged to its present form in less than a century without a fundamental backbone. There's not one universally accepted set of procedures and rules that have guided plastic pollution research. Instead, these standards are being developed by Kristian and his contemporaries, who have been forced to get creative, detecting and collecting microplastic that exists in the air, land, and sea with all manner of environmental research equipment—often novel, jerry-rigged, or repurposed. Aboard *Christianshavn*, for example, we would carry out our plastic experiments with devices typically used to collect plankton and water samples. This, in an effort to understand the actual scale and scope of microplastic pollution in the oceans.[12]

Some scientists have developed homemade contraptions that non-scientists can use to gather data on microplastic in the natural environment. Max Liboiron of the Civic Laboratory for Environmental Action Research (CLEAR) at Memorial University of Newfoundland, for example, has developed a research device they call BabyLegs, which, once easily assembled using common household items—including nylon tights for infants and empty soda bottles—can be dragged as a trawl across the surface of any body of water to catch particles of microplastic.[13] In fact, a significant amount of data on the fate of plastic in nature has been gleaned by so-called community scientists, those among us who lack a professional or academic background in science but are interested in helping the scientific world by collecting information about their local environment. Much community science data on plastic is open source, meaning it's been made public online for anyone to see, use, and contribute to.[14]

By the power of invention, the art of navigation, and the tenacity of spirit, humans have sailed the seven seas and have sent high-tech equipment—and even people, in pressurized deep-sea submersibles—to the lowest point on the planet. Collectively, our almost eight billion sets of human eyes have seen just 20 percent of the oceans.[15] But just because not much microplastic is visible at sea doesn't mean there's not much there. Microplastic is most certainly present in the sea, at all times, and at a staggeringly high count of at least fifty-one trillion pieces throughout the world's oceans, according to United Nations estimates. That's a count five hundred times greater than the number of stars in the Milky Way galaxy, the colossal cosmic neighborhood Earth calls home.[16]

The very smallest pieces of microplastic on our planet, which some scientists say ranges in size from one to one thousand nanometers in diameter, the size of a single bacterium or smaller, have another newly coined moniker: nanoplastic. These bits of plastic are so small that they often slip right through nets designed to catch microplastic, which is slightly larger in size. Scientists have not even begun to estimate how much nanoplastic exists in the oceans, because it's so difficult to identify and quantify, though those who are now looking for it have a hunch that there is quite a lot out there. As the ocean's existing load of plastic items and microplastic continues to break up, there will only be more nanoplastic in the sea over time.[17]

To approximate the total number of plastic pieces currently circulating the surface waters of the eastern North Pacific Gyre, Kristian would have to count each and every particle of microplastic captured on the ship's numerous trawl runs back at his lab at Roskilde University in Denmark and, using the coordinates taken when each trawl was sent out and picked up, average the number of particles found over this area of the gyre. He'd also classify—chemically and by eye—what kinds of plastic the crew found, for the plastic items we make and use are derived from so many different chemical combinations.[18]

"It's a daunting task," said Kristian, who had agreed to help analyze the microplastic samples Plastic Change collected in the Pacific and on its most previous expeditions through the Atlantic and Caribbean. "Hopefully, I will get some graduate students to help me soon," he added with a half smile, considering both the logistical magnitude and scientific importance of the job placed before him.

However tedious, counting microplastic pieces pulled from surface waters has been the most popular method for estimating microplastic concentrations in any water body—fresh, salt, or in between—for the past five decades. In the oceans, tiny plastic pieces often remain well camouflaged in the peaks and troughs of each wave that ripples or crashes. Like flecks of gold concealed by a rushing stream until they are patiently panned out, the microplastic and nanoplastic bits permeating the oceans largely remain invisible until a net is dipped into the water. Humanity's monitoring of the quantity of microplastic in the ocean's surface waters with trawls—an efficient and accessible research tool—is a large and ongoing effort taken up by expert and amateur scientists alike. It gives on-the-water insight into how much plastic is getting into the sea.[19]

All hands are now on deck—and we're going to need them.

"In a little more than sixty years, we know we've littered more than 150 million metric tons of plastic into the oceans," Henrik said, sitting in the cockpit and marveling over the collection of plastic debris he and the crew had nabbed out of the gyre.[20] "When we trawl [the surface], we find less than 1 percent of what we litter throughout history. So the main question we're trying to answer with this expedition is, Where does it all end up? Is it in the fish? Is it in the birds? Is it on the beach? Is it on the deep-sea floor? Where has all the plastic gone? Is it in us humans?"

I'd later learn that, while it's become the de facto method of studying plastic pollution, counting microplastic collected from the surface alone tells scientists a limited amount about the total plastic load of the oceans, as well as other water bodies like lakes, rivers, and streams. Wind and sea conditions are capable of both attracting plastic to the surface and diverting it below, creating an inconsistent sampling area from which scientists have had to perform their research. But, as Henrik noted, something is not adding up: According to plastic production and municipal waste data, there should be more plastic in the sea than surface trawling is telling scientists. In other words, the best available science may be leading researchers to underestimate how much plastic is in the sea—possibly by millions of tons.[21]

Experts are working to piece together an understanding of where all our plastic is going if it's not degrading, as well as what it's doing—to our environment and to us.

# CHAPTER 3
# The Ocean's Canaries

In this predawn hour, the ship was surrounded by complete darkness, save for a few stars in the sky and the streaks and sparkles of atomic-green water that followed it, a trail of bioluminescent algae—simple plant-like organisms that glow brightly when disturbed by water movement, making them most visible at night.[1] Of all sailing shifts, the 4:00 a.m. to 8:00 a.m. spot is the most grueling. It requires an intentional hijacking of the natural diurnal human biology in order to steer, navigate, hoist sails, tie knots, winch lines, and just generally stay awake during nocturnal hours.

The wind was light, the sailing slow. I slumped along the starboard rail, fighting off the urge to nod off by listening to the most earsplitting music I could stand. I extended an earbud to Rasmus—who sat silently at the wheel, sipping tepid, muddy looking coffee from a chipped ceramic mug—sharing a song only we could hear. Frequent doses of coffee and licorice, and loud music, usually did the trick of staving off the temptation of rest. Though the possibility of running into additional human company was slim, it could happen, and one always needed to stay vigilant. In fact, the previous day we had crossed paths with a hulking Matson cargo ship heaped with improbably tall towers of white steel containers, some stacked six high. The ship was heading in the same direction—west, toward Hawai'i—though she quickly outpaced us.

In these somnolent early hours, my mind slipped into surreal dreams. It was easy to imagine the entire world was the sea and our ship filled with the last people on Earth.

I was shocked out of my reverie by the subtle intrusion of light in the darkness: The sun had begun its ascent up into the sky, a faint yellow glow illuminating the slim cracks in a sky full of cumulonimbus clouds. Black waves below began to shine and quiver until the whole sea was like quicksilver, reflecting ripples of new morning light.

Chris was first of the crew to emerge from below deck, just catching the tail end of the sunrise. He silently scanned our surroundings, as he did every morning, camera in hand. "Majestic," he whispered to no one in particular, clicking a few dozen frames of the silver sea before disappearing through the hatch, presumably to get a bit more sleep before his shift. Before long, Malene, Henrik, and Torsten were sitting with us in the cockpit, munching through bowls of plain muesli, discussing the day's experiments. At present, *Christianshavn* was moving at a very slow clip of three knots per hour, perfect conditions for trawling the surface.

Malene and Kristian knelt on the bow, tying a flat, boxy metal contraption with a long net "tail" open in the front and closed in the back to the ship's spinnaker pole, a metal boom designed to hold an extra sail set forward of the mainsail. The spinnaker is often useful for catching wind when it blows from behind the ship. On this trip the spinnaker pole would be repurposed to drag the research device across the ocean's surface, adjacent to the ship.

The device, called a manta trawl, is named to honor its rough likeness, in size and shape, to the flat-winged, horizontal-swimming creature we call a manta ray. Manta trawls aren't meant to collect the larger plastic items—bags, bottles, containers, ropes—floating in the ocean. Instead, their slim mouths and fine mesh nets are specifically designed to scoop up microplastic. And, as we know, while a surface-level trawl alone can't tell us how much plastic is in the ocean, it's proven a critical first step to estimating pollution levels.[2]

"Tre, to, en, nu!" Kristian counted down to the toss. As Henrik stood at the helm, Malene and Torsten hefted the manta trawl into the water. The waves were considerably calmer here, gentle humps instead of massive, frothing swells. Above, cloudy skies began to clear. The glinting aluminum device bucked, splashed, and dove as it settled into cruising mode on the sea surface, dragging alongside the ship.

Four hours later, around two o'clock that afternoon, Peter took the wheel and nosed *Christianshavn* into the wind as Malene let out the mainsheet. The mainsail relaxed before rippling like the ocean's surface, luffing in the light wind. Up on the bow, Kristian cast a long gaff at the taut trawling rope. First a miss, then another miss, and then a snag. With Torsten's help, he reeled in the unwieldy device from the water. Once the pair pulled the trawl onto the deck, Kristian inspected its long tube-like net. The tight, opaque mesh dripped and glistened in the afternoon sunlight. It was filled with colorful, confetti-like bits, likely plastic, Kristian remarked, as he held the net up to his wide blue eyes for a close-up look.

"Look at that! Look at that!" Chris marveled at the plump net.

"Wow, so much plastic!" Sofie remarked.

Sensing the crew's excitement, Kristian pointed out that manta trawls also often scoop up biological material, such as algae, jellyfish, kelp, fish larvae, and fish eggs—in addition to plastic. That could account for some of the trawl's bulk. These things can make it more difficult to determine, with a quick look, roughly how much microplastic—which is similarly shaped and colored—has been collected. Yet Kristian, an experienced plastic researcher, spoke with certainty as he poked at some of the more easily identifiable pieces inside: several soft white balls of expanded polystyrene, black nylon threads, and blue shards of hard high-density polyethylene.

Kristian unscrewed the trawl's net tail from its body and passed it to Malene, who sat in the ship's teakwood cockpit. Piled next to her were three metal sieves with mesh sizes of 0.3, 1.0 and 5.0 millimeters, a few pairs of metal pincers, a dropper bottle of water, a notebook, a pencil, and a large box of glass sample jars. Kristian settled down next to her and stacked the three sieves with the smallest-meshed sieve on the bottom and largest-meshed sieve on top. Malene emptied the net onto the stack, coaxing each bit of microplastic out with gentle splashes of freshwater dispensed from the dropper bottle.

The bits of microplastic settled, the largest pieces sticking to the top sieve while the smallest pieces got snagged on the bottom sieve. The medium-sized microplastic pieces were caught in between. Separating the various-sized pieces of plastic into different sample jars on the ship would make the job of analyzing them in Kristian's lab in Denmark a little more organized. Malene had marked her notebook with the date, trawl time out and in, and the ship coordinates at the start and end of the procedure.

After thoroughly emptying the net of its microplastic catch, Malene separated the sieves. Each was coated with a colorful spread of microplastic pieces—some jagged, some round, some smaller than a sesame seed, some larger than a nickel. All appeared and felt weathered and brittle with a matte texture, their characteristic plastic luster lost from time spent being tossed silently around the sea. Malene carefully rinsed the contents of each sieve into small amber glass jars for safekeeping. The sailors would see to collecting as many manta trawl samples as possible during the journey across this region of the gyre.

Such labor-intensive work is necessary because bits of microplastic and nanoplastic look and feel far different from the plastic items they have broken off of. In the ocean they become brittle, worn, and attractive to chemicals—which bond to their surfaces. Pieces of microplastic and nanoplastic that have been submerged in ocean water act like small sponges, absorbing toxins—such as pesticides and heavy metals—from seawater. What's more, many plastic items are manufactured with added chemicals, called plasticizers, which change the structural properties of the material—making it stronger, more heat resistant, or more malleable, for instance. Plasticizers are part of what makes the difference between a sheet of plastic cling wrap and a plastic water bottle. And so, each piece of plastic in nature has a story to tell that can help us gauge its danger: Out in the ocean the microplastic bits may look like anonymous polymers. But in fact, as I'd later learn upon reaching dry land, each piece of plastic possesses a chemical formulation that can reveal much about its origins—and dangers.[3]

Relief from the sun's rays became more valuable the closer we sailed to Hawai'i; as we dipped south, arcing toward the island chain, the sun's intensity, heat, and reflection off the sea reached an overwhelming level within just a few hours following the sunrise. Around midday, my temples dripping, I tucked myself into one corner of the wooden bench at the stern to write in some modicum of shade beneath the ship's solar array. Rasmus assumed his usual seat next to me, headphones over his ears and his fingers dancing along the brim of his straw hat. His eyes monitored the poles he'd cast, sitting in their chrome rod holders. Suddenly, one of the poles lurched forward and began unspooling meters and meters of line from its whizzing reel. Rasmus spun around and grabbed hold of his fishing pole, reeling in the creature on the other end of the line slowly and carefully. I leapt off the bench, out of the way. The sailors' heads turned toward the jolting sound and the wide expanse of calm sapphire water astern.

A big shining fish leapt out of the water in a wide arc, sending a dazzling spray of seawater through the air. "It is a mahi-mahi!" exclaimed Torsten, watching the animal's amorphous, atomic body, electric yellow-green and speckled with fluorescent blue, dance above the water. "This is the most beautiful fish in the world," he marveled, as he often did, in awe of the nonhuman creatures who happened to intercept us.

This would be one of eight mahi-mahis we would catch and eat during our twenty-three-day expedition. Rasmus and Peter had been taking turns reeling them in. The crew grew more elated as Rasmus wrested the fish nearer the ship. Finally, Peter dipped down a net and pulled the fish across the threshold separating sea from ship. At that point, the sailors emitted a few cheers before moving on with the day's tasks, knowing then we'd have a fresh fish for dinner. But Chris and I kept our eyes fixed on the scene. Being connected to wildlife in our own ways—Chris as a documentarian of albatrosses and me as a licensed wildlife rehabilitator working to heal sick and injured owls, turtles, and other creatures—we both experienced pain when watching a nonhuman life slip away. Call it empathy, call it realism: We both felt it important to bear witness to this aspect of eating other animals whenever possible.

Chris and I resisted averting our eyes, though we both shivered, when Rasmus spiked a screwdriver into the fish's skull with one hand, keeping the fingers of his other hand wrapped around the fish's slick, thick body. Blood spilled out of the point of impact and down the fish's face as dark crimson tears, while the fish's entire body shook, squirmed, and then shuddered until it ceased moving. Chris stood next to me, weeping, both of his tattooed hands over his dark brown eyes. He said a prayer for the fish and excused himself to his bunk.

My mind began to drift toward the possibility that the fish might contain microplastic. Apparently, so had Kristian's. The scientist ambled across the deck from the bow, where he was busy with the manta trawl, to the stern, to take a look at the fish. He pointed out that there are few, if any, ocean animals today that have *not* eaten plastic at some point in their lives. With that in mind, I watched expectantly as Rasmus pressed a shining silver blade into the mahi's plump belly, waiting to see what might emerge.

Rasmus opened the mahi-mahi's taut, protruding belly and used his bloodied fingers to extract a tangle of purplish-pink innards and throw all of them—save for the stomach, which he dropped into an empty container—overboard.

Kristian took the container and held it out to me for inspection. I prodded the slimy stomach with my finger and inhaled the metallic scent of spilled blood. He explained it was likely we'd find some micro-plastic inside. My eyes flitted from the contents of the container in my hand to the fish itself, which was nearly unrecognizable from the ani-mal that was, just a little while ago, so full of life and color—and not just because Rasmus had disemboweled and started slicing her up into pieces. As soon as it is hooked, a mahi-mahi undergoes a drastic trans-formation: The process began in the water during the fish's fight for life. The fish's flashy greens, blues, and yellows began fading slowly, at first by nearly imperceptible degrees. By the time Rasmus exhausted the fish and pulled her up onto the boat, the colors of her skin, while still bright, had dulled significantly. After Rasmus knocked the life out of her, the mahi-mahi's colors rapidly dulled and grayed.

When Kristian moved to dissect the mahi-mahi's stomach, Rasmus turned to me, his co-chef, to consult about our dinner preparations. "I'm thinking in the oven with oil, garlic, salt, pepper, and lemon," said Rasmus, his sea-blue eyes shining.

I smiled, for his sake, but internally felt conflicted. The idea of eating yet another fish caught in what many consider to be the most plastic-polluted stretch of ocean in the world was starting to make me uncomfortable.

Just as I was about to weigh in on the dinner deliberations, I felt a tug on my shirttail. "Uh, Erica?" It was Kristian, sitting behind me, the slick pink mahi-mahi stomach opened up on a wooden cutting board situated by his feet. The still slightly seasick scientist wiped a bead of sweat from his brow with the back of one hand and with the other reached inside the stomach to extract a small, intact blue-silver fish. It possessed fanlike transparent appendages more reminiscent of a bird's wings than a sea creature's fins.

"En flyvefisk—a flying fish," Kristian explained. "I'm going to cut it open."

Malene and I gathered around Kristian to watch as he sliced down the fish's belly and pulled out its tiny stomach. He pierced the plump little organ with the knife and squeezed. A viscous, pink-tinted liquid oozed out, carrying with it what looked like a cluster of translucent fish eggs. All but two were orange tinted. Kristian grabbed the two dull-looking outliers with his fingers and pressed down. They did not yield to his weight; they didn't burst like a fish egg or bit of plankton would. Instead, they felt hard and rigid, and their shape was much too perfectly spherical to be a product of nature.

"Plastic! I think we have plastic," Kristian exclaimed.

Now I felt even more uneasy about this whole fish-catching—and eating—endeavor.

Kristian showed us two suspected plastic spheres, commonly called *nurdles* or *pellets*, which are produced industrially for melting down to form plastic items.[4] He reiterated that, because microplastic is so prevalent throughout the oceans today, it's eaten by all kinds of aquatic organisms from the smallest planktons to the largest whales and, by extension, people too.

Fishing is among the simplest ways to procure food when you find yourself on a sailboat. And it is an activity engrained in human culture, a part of the rhythm of daily life—work, recreation, and nourishment—in coastal regions like Long Island, where I grew up. Yet for many people around the world, fishing is not just a part of their culture but a crucial part of their survival.

According to the Food and Agriculture Organization of the United Nations, there are about 3.2 billion people around the world who rely on wild-caught seafood to provide them with a significant amount— at least 20 percent—of their annual protein intake.[5] With the world's human population now surpassing 7.7 billion, and expected to surge past 9 billion by 2050, the need to address global food insecurity— which presently affects at least 820 million people globally—is becoming more urgent.[6]

Fish, and other marine animals, feed enormous numbers of people. In developing coastal regions and on small islands, particularly in Asia and Africa, seafood can comprise 50 percent or more of a person's animal protein intake. People in these regions, who are often food insecure, would have trouble surviving if they could not catch and eat fish.[7]

Fish also feed enormous numbers of wild animals, not only those living primarily or exclusively in lakes, rivers, and oceans—like whales, dolphins, squid, sharks, octopi, otters, sea lions, water snakes, eels, and other fish—but also those who fly above it, such as osprey, eagles, boobies, frigatebirds, albatrosses, terns, and cormorants. Even many creatures dwelling on shorelines—like bears, wolves, and jaguars—fill the better part of their bellies with fish on a regular basis.

Plants, too, are fed by fish: When fish die and decompose, the nitrogen, phosphorus, fat, and protein held by their bodies are released to enrich the soils from which florae propagate. This fact has been famously demonstrated in the Pacific Northwest and Alaska, where scientists have traced a specific type of nitrogen found only in the oceans to the bodies of decaying salmon, and eventually to towering cedars and other trees.[8]

"Salmon not only help grow the trees, but they actually grow in the trees," photographer and journalist Amy Gulick wrote in her book

*Salmon in the Trees: Life in Alaska's Tongass Rain Forest.* "Once you understand this remarkable connection you quickly see that everything is connected and we need every link in this web of life in order for the whole place to function."[9]

What could it mean for the survival of all life on Earth if fish are becoming increasingly filled with plastic? In 2015, experts estimated the amount of plastic in the oceans would outweigh fish by the year 2050.[10] We're on track to get there, if we're not there already: By 2020, humans had created enough petrochemical-based plastic to outweigh the mass of all marine *and* land animals combined, by a factor of two, according to scientists at the Weizmann Institute of Science in Israel.[11]

In the oceans, plastic acts as an imposter. A transparent plastic grocery bag floating upside down near the surface looks a lot like a slow bobbing jellyfish to a hungry leatherback sea turtle. Aggregations of tiny blue microplastic particles stuck to a piece of kelp may register as a cluster of plankton to foraging baby fish. Fragments of clear and light-colored microplastic can seem like small fish, deceiving young mahi-mahi searching the seas for food.[12]

In most cases, we assume animals eat plastic because they misidentify it as another food. Many animals, especially those who dwell in extreme and dynamic environments like the oceans, rely on a body equipped with a suite of fine-tuned senses to locate the prey or plants they eat. Most marine species' predominant navigating sense shifts in tandem with the ever-changing ocean conditions they encounter. Plastic can clearly look like other food sources in the ocean, but not all ocean animals use sight to find what they need to eat. Many far-ranging marine animals—including some fish, sea turtles, and seabirds like shearwaters and petrels—use their sharp sense of smell to find food when traveling thousands of miles across the ocean.[13]

There's also a chemical element to marine animals' proclivity for eating plastic. A commonly shared food source among ocean predators is zooplankton—tiny marine animals that consume algae, simple marine plants at the base of the ocean food web. When zooplankton graze on algae, the algae release a chemical called dimethyl sulfide (DMS) that smells appetizing to zooplankton. Many land plants also emit odiferous chemicals upon being eaten that act to attract other hungry animals. Aware that some seabird species rely on DMS tracking to find food at sea, Stanford University postdoctoral research fellow Matthew Savoca has studied whether microplastic might be sending a similar signal. Using chemical analyses, Savoca detected high concentrations of DMS on weathered, algae-coated bits of microplastic scooped from the sea. Microplastic's adoption of a strong DMS odor seems to have created "an olfactory trap," as Savoca put it: "These results suggest that plastic debris may be more confusing and appetizing to marine organisms than previously thought possible."[14] Scientists estimate that 90 percent of the world's seabirds have ingested plastic at some point in their lives. By 2050 they expect 99 percent of all seabirds will have eaten plastic.[15]

Savoca has also focused his research on the anchovy, a tiny fish and another plankton-eating animal with enormous ecological importance. After plankton, anchovies constitute a next step up in the marine food web. They are "forage fish," small schooling fish hunted in droves by many larger marine predators.[16] And alarmingly, the anchovy—which is essential to so many animals' diets—is one of many marine animals that appear to feast on microplastic because they are attracted to its DMS scent. Because microplastic has been shown to move up the food web, Savoca and his collaborators have warned that anchovies' and other forage fish's predilection for plastic could be adversely affecting other marine animals, and possibly even human health.[17]

Corals also have displayed an affinity for microplastic, at least in lab experiments. Like seabirds and fish, corals seem to be selectively eating small plastic particles because plastic's chemical signature is attractive to the sensory systems they use to discern what is food. Scientists at Duke

University, led by PhD student Austin S. Allen, have found that corals won't eat little bits of sand—though they're the same size as tasty zooplankton—but that the corals *will* readily eat little bits of similarly sized plastic, which also look like zooplankton.[18]

DMS or another chemical cue seems to be stimulating corals' appetite for plastic, and at what appears a significant cost: In Germany, scientists at Justus Liebig University Giessen have also found laboratory evidence linking corals' ingestion of and other contact with microplastic to bleaching and other signs of disease like increased mucus production, tissue necrosis, algae overgrowth—and ultimately, death. This, even if corals succeed in spitting the microplastic out of their bodies following ingestion. It could be the plastic that is interfering with corals' bodily functions, or the chemicals the plastic is carrying, or both.[19]

Corals cover just 1 percent of Earth's surface, yet they're estimated to provide the economy with an outsized boost of $29.8 billion per year, around the same GDP as a small country. Reefs pull in tourists and recreationists who spend $9.6 billion on travel, equipment, and activities; provide $9.0 billion in coastal protection as a natural buffer from storms; offer a home to often-fished marine species, pushing $5.7 billion into fisheries; and support more than a million marine plant and animal species, the research and conservation of which injects an estimated $5.5 billion into the global economy annually.[20]

Despite all they give people, corals are rarely recognized as the clever and resilient animals they are. When corals reemerged after their first extinction event a few hundred million years ago, they forged an important mutually beneficial relationship to algae called zooxanthellae. These are the tiny organisms that coat the exterior of corals, lending brilliant colors. Like other kinds of ocean-dwelling algae, zooxanthellae convert sunlight into energy and produce oxygen from carbon dioxide. Corals, which supply their colorful algal allies with carbon dioxide and physical protection, use the energy and oxygen generated by the algae to reproduce, grow, and maintain the strength of their calcium carbonate skeletons. Corals may also eat drifting zooplankton, which they stun and catch with tiny barbed harpoons built into their bodies.[21]

In the best situations, when corals are thriving and reproducing regularly, they cobble into colorful, crowded, colossal reefs capable of supporting a rich menagerie of creatures attracted to the tantalizing offer of both nutrition and protection. About a quarter of all marine animals live on and around reefs, "the rainforests of the sea," in great variety, including myriad fish, rays, sharks, sea turtles, marine mammals, and plants—and the oceans as a whole rely on reefs to stay in balance.[22]

When sea temperatures soar too high, corals revolt, and this can quickly clear entire reef ecosystems of life in a way that's reminiscent of a classic boom-town-turned-ghost-town scenario. Healthy coral polyps grow agitated in extreme warmth, squirming and squeezing, spitting the symbiotic algae out of their bodies into the surrounding seawater. With the algae goes corals' color. Stressed-out polyps' bare bodies fade like pigmented paper left out in the sun and eventually turn a stark and skeletal white if left continuously exposed to high temperatures. This process, appropriately called bleaching, renders corals weaker and more susceptible to diseases.[23]

Worldwide, corals are in crisis. As ocean temperatures have warmed as a result of climate change, waters have also become increasingly acidic —a product of so much extra carbon, poured into the atmosphere from our combustion of fossil fuels, dissolving into the sea. Acidic waters cause coral skeletons to disintegrate and make corals' necessary bone-building process—especially during a post-bleaching repair—an extremely uphill climb. Corals can recover from bleaching but face an increasingly slim chance of recovery as the oceans continue to acidify. If exposed to too-warm waters for a long period of time, a coral may never regain its symbiotic algae, may never regrow its skeleton, will never recover—and as a result, will surely die.[24]

The late Ruth Gates, one of the world's foremost reef experts, once described corals to me as "the ocean's canaries: highly perceptive and highly vulnerable creatures." There's plastic pollution, and then there's the overwhelming burden of climate change, as well as sediment from overdeveloped lands crumbling into the ocean, nutrient buildup from sewage discharge pipes routed to the ocean, fertilizer and toxic pesticide runoff, and oil leaks from steadily increasing ship traffic and deep-sea wells. Add it all up, and corals' chances of survival into the future appear absolutely dismal. Losing corals is a major concern for the health of the oceans, and in fact the entire planet.[25]

One morning while in the gyre, we had encountered several charcoal-colored, black-footed albatrosses, each equipped with a seven-foot wingspan slung like a recurve bow across a warrior's back. The magnificent seabirds dipped in and out of the blue waves in search of squid, krill, and fish eggs. Torsten slipped into the hull to wake Chris, who would surely not want to miss what was happening outside. Soon, one sailor had told another until the entire crew was gathered up on deck, shoulder to shoulder, entranced by the albatrosses.

Albatrosses are one variety of seabird that seems especially attracted to eating plastic particles and objects. They are experiencing widespread mortality as a result, as the whole crew was aware, thanks to Chris Jordan's aforementioned documentation of plastic's toll on albatrosses on Midway Atoll, a remote coral ring encircling three islands still haunted by the ghosts of combat. On these tiny Pacific islands where Allied soldiers once trained and fought during World War II, today one may find a few dozen caretakers and researchers patrolling the perimeters of a mostly deserted former military base. The few people who are either employed by the US Fish and Wildlife Service, or secure the federal permits necessary to spend time on Midway, like Chris, are there mainly for the birds.[26]

Each year, come November, the atoll comes alive when avian migrants who spend most of their lives in near-perpetual flight over the ocean touch down on solid ground to briefly copulate, lay eggs, and raise babies. For about half a year, more than twenty-three species of ocean-faring birds blanket the atoll's total landmass of just under two and a half square miles, across three islands: Sand, Spit, and Eastern. The ranks of these feathered travelers include about a million adult albatrosses—mostly Laysan and black-footed albatrosses, but occasionally a few short-tailed albatrosses—and, by January or February, their newly hatched chicks. The albatross babes—plump, fuzzy, and flightless—cannot procure their own food. Albatross parents must take turns flying out to sea so their families can eat.[27]

Chris visited Midway twice in the first decade of the 2000s, capturing images he would ultimately incorporate into his 2009 photography series of the dead albatrosses, *Midway: Message from the Gyre*, and his film *Albatross*. Though it had been several years since he captured his footage, when we went sailing, he was still working on distilling the sometimes sixty-plus-year lives of albatrosses into a film of a palatable length.

The day we saw the albatrosses, Chris propped up his laptop inside his bunk, top and center in the saloon. Half the crew, those not already at or preparing to take the helm, perched themselves atop the wobbly saloon table to watch Midway's albatrosses in the ship's makeshift theater. While waiting for their parents to deliver food in open, twig-lined nests, the young albatross chicks dotted the low-lying islands' grassy interiors like a bloom of great gray flowers amid a tangle of plastic debris. Around them, crumbling concrete guard boxes, hollowed-out skeletal steel buildings, and rusted-out war equipment continued their long deterioration into the sand and a rising sea. Off screen, out at sea, albatross parents were busy filling their stomachs with squid, krill, and fish eggs floating on the water's surface, where plastic items and microplastic also lurk.

One by one, each albatross parent swooped back onto the shore and then set upon the task of finding *their* chick among so many other seemingly identical peeping balls of fluffy gray down. Chris's camera zoomed in on one returning parent, a black-and-white Laysan albatross, who waddled over to their hungry offspring. The chick emitted a few peeps and began to tap on the slender yellow-pink curve of their parent's beak, propelling their caregiver to regurgitate what they had swallowed at sea. Along with seafood, the albatross parent coughed up a colorful stream of microplastic bits—clearly visible through Chris's lens—into their chick's gaping mouth.

The film ran on: Chris encountered a series of young albatrosses, clearly dying, by the shore, on a cement ruin, in a nest, on the grass. Writhing on their sides, mouths agape, all seemingly choking on something unseen, but not unknown. In the film, Chris concedes there is no way he can stop the life from slipping out of these albatrosses. A snip of a scissor through the belly of one of the freshly dead birds sent the obvious cause of death tumbling out: disposable lighters, microplastic fragments, wrappers, straws, bottle caps, children's toys, and other plastic things—nothing even vaguely resembling the squid, krill, or fish eggs albatrosses are designed by nature to eat. Chris makes his rounds, spilling the seabirds' plasticized guts out all over Midway.

As heart wrenching as some of *Albatross*'s images were, still the sailors watched the lives and deaths of the film's namesake seabirds unfold onscreen. Late summer came, and the formerly flightless babies—the lucky ones with lives not prematurely taken by plastic, sea level rise, invasive mice and rats, or other human-introduced hazards—finally shed the peculiar fuzz of their youth to bear the sleek plumage worn by their parents. One by one, these young survivors took to the skies. They set off toward the sea in a wide-legged run, sending sand spraying in all directions, and finally, just before it seemed they would crash into the ocean waves, made one final swift kick up to catch the wind. Up they went. In the sky, this new realm, they folded up their legs and embarked on a first flight, the duration of which would extend three to five years over this open sea.

Today Midway possesses a meager human population. But the relics of humanity are everywhere: Plastic items and microplastic arrive from far and wide, creating disastrous living conditions for the atoll's numerous avian inhabitants. A steady, unyielding stream of plastic debris washes onto birds' nesting grounds across the atoll and circulates in proximal Pacific waters where a variety of albatrosses and an assortment of other seabird species forage.

While some marine animals like seabirds, fish, and corals may be attracted to eating plastic because it tricks them into believing it is food, other animals may not be aware or able to control whether or not they ingest it. This is the case for filter-feeding animals of all sizes, from the blue whale, Earth's largest living animal, which can grow as long as three full-size school buses parked in a straight line, to blue mussels, which can fit comfortably in the palm of a human hand. These very different creatures both use their bodies to sieve small living organisms like krill and plankton out of large quantities of seawater. Because microplastic is roughly the same size and shape as filter feeders' food sources, it is easily—though unintentionally—incorporated into their diets.[28]

I'd gain some firsthand insight into the ubiquity of unintentional microplastic consumption in the seas a few years after Plastic Change's Los Angeles to Honolulu expedition, when Torsten and I sailed together again, this time around the western and north coast of Iceland. On that journey, arranged by Ocean Missions, a marine conservation organization newly founded by Spanish marine biologist Belén García Ovide, we observed humpback, orca, minke, and sperm whales feeding in waters carrying considerable amounts of microplastic. Hers would be among the first groups in Iceland to investigate local microplastic pollution—despite the country's emergence as a tourism destination following the dramatic eruption of its Eyjafjallajökull volcano in 2010, and all the plastic the influx of visitors inevitably brought.[29]

While the Icelandic waters were not as polluted as the eastern North Pacific Gyre, they certainly contained microplastic. The remote, frigid sea was also full of life, teeming with cod and capelin; and overall, it seemed our manta trawls were more likely to swallow seaweed, plankton, or fish eggs—favorite foods of marine creatures—than microplastic. However, this also meant microplastic was mixing with the whales' food, right where we'd witnessed whales actively searching for something to eat.

I remember watching a small group of humpbacks, their barnacled, pleated chins swelling like bellows as they gushed great mouthfuls of seawater through enormous baleen-plated jaws, entrapping krill, capelin, copepods, and small fish. A filter-feeding adult humpback whale draws in nearly nineteen thousand liters of water and one and a half metric tons of krill per day. As the crew observed the whales, Canadian marine biologist Charla Basran, who'd also joined Ocean Missions' expedition, had remarked, "Wow. Imagine how much plastic they trap in each gulp . . ."

Our trawls revealed that where sea life gathered, microplastic was also likely to intrude. The whales really could do nothing to avoid eating it.

Surprisingly little research has been done to understand why microplastics and wildlife are congregating in the same areas of the ocean. However, scientists affiliated with the University of Siena in Italy have noted that in the northwestern Mediterranean, cool, nutrient-rich offshore currents appear to attract both plankton and microplastics into a key fin whale feeding ground, the Pelagos Sanctuary for Mediterranean Marine Mammals. Like humpbacks, fin whales use mouths outfitted with baleen to filter plankton and small fish from seawater. If microplastic is aggregating in whales' feeding grounds, it's likely they're consuming it.[30]

Basran, who studies whales in Iceland, pointed out that the deep seas we were sailing also experienced much upwelling of cold, nutrient-rich water. Off Iceland's northern coast, an unusual confluence of warm Atlantic and cold Arctic currents also seems to attract marine life—and people.

"Where there's plankton and fish, whales and fishermen follow," Basran had said, pointing to several hulking fishing vessels looming nearby, spooling in their plastic longlines. From these Icelandic waters, we pulled out tiny nylon fibers, which we could clearly identify as pieces of fishing nets and lines. Like the gear being hauled up by the fishing boats, these lines were clear and blue.

All across the world, nonhuman animals are suffering the consequences of humanity's plastic addiction far more quickly than we are learning how exactly they—and perhaps, also, we—are being harmed. What's more, all the microplastic ingested by or entangling animals—in addition to that skimmed off the top layer of Earth's waterways—does not account for all the plastic that's expected to be in the oceans. To find the rest of humanity's "missing plastic," scientists have had to dive deeper—probing for plastic out of sight, below the surface.

# CHAPTER 4
# From Ship to Shore

Bleary eyed, I rolled out of my bunk one morning, bladder near bursting, and took the two necessary short steps to the ship's toilet, which had been installed in a small closet near my bunk. Slapped across the door was a blue strip of painter's tape onto which the toilet's most previous occupant had scrawled the words "do not use" in black marker, for some yet-unknown reason. I snapped on my life vest and headed on deck where I yawned good morning to Henrik, Kristian, Malene, and Torsten. They were sitting in the cockpit poring over an array of plastic sampling nets and tubes.

Without having to ask why I had come up, or what I needed, Malene handed me the blue bucket marked "TOILET," which we kept on deck. The men had it easier, as they could quickly relieve themselves over a rail without squatting over the bucket. Before we'd set off from Los Angeles, during a crew meeting, Torsten had reminded all the men on board to clip their life vest to the ship each time nature called. "Most drowned men are found with their flies zipped down," he remarked. I thought about this, and laughed, as I made my way to the bow where I plunked the toilet bucket down against the scant privacy of the ship's brass mast, silently willing any rogue waves away. It is hard enough to balance your ass on the rim of a plastic bucket under normal conditions. To do so without skittering, along with the bucket's contents, across the deck of a moving ship when a wave hits requires considerable concentration. Even still, if you managed not to fall over, it wasn't unusual to get soaked by sea spray while sitting "behind the mast," or when emptying and rinsing the bucket over the side of the ship in preparation for its next user. All that to answer nature's call.

When I returned to the cockpit, the science crew—Henrik, Kristian, Malene, and Torsten—were rigging up their latest plastic sampling contraption, a vertical trawl. The device comprised eleven tubes, each backed on one end with a net, strung down a twenty-meter line. The top end was tied to a buoy that would sit on the ocean's surface, while the bottom end was weighted down by a light anchor that would hold the line taut below. The contraption looked unwieldy to say the least.

Like the manta trawl, the vertical trawl was tied to and dragged from the spinnaker pole, holding it at a distance to prevent it from scraping up the side of the ship. Henrik tossed the vertical trawl in the sea where it unfurled, and we watched as, one by one, each tube slipped beneath the waves. According to Kristian, it was one of the earliest instances such a device had been used to look for microplastic below the surface of the Pacific Ocean.

After a few hours of dragging the trawl, Rasmus, who was at the wheel, directed *Christianshavn* into the wind, sending the sails luffing, halting the ship to a stop. Kristian and Torsten hefted the trawl up on deck, and then we continued on our way. In the cockpit, Malene and Henrik removed the eleven nets and emptied the solid contents of each into labeled glass sample jars. The jars were filled with tiny particles, indicating the trawl had indeed done its job of catching things, though it was as yet unclear whether or not those things were plastic. Given what we had witnessed over the past few weeks at sea, however, it seemed like a good bet that they were.

During the day, the uppermost depths of the remote Pacific are relatively quiet. Much more happens when night falls over the ocean. The darkness heralds Earth's largest migration, when fish, plankton, squid, crustaceans, and other creatures who spend their days near the seafloor or in the middle layers of the ocean swim up hundreds to thousands of feet to the surface to feed. These migrating sea creatures play a critical role in cycling nutrients throughout the marine food web. When the sun rises, they return to a greater depth, where they defecate, and their feces rain downward, carrying nutrients to the relatively unproductive seafloor. This fertilizes it and makes it habitable for a great many creatures.[1]

With the vertical trawl, we were looking for any evidence of microplastic throughout the ocean's uppermost layer of water, where so many migratory ocean animals come to feed at night. If there was plastic, these ecologically important animals were probably eating it—and it could be impairing their survival. What's more, presence of plastic just below the waterline could indicate a significant amount of plastic overlooked by surface manta trawling equipment. If there's microplastic in this region of the sea, it probably means scientists have been seriously underestimating the ocean's total plastic pollution load as expected.[2]

Later that day, as the sun disappeared somewhere beyond the infinite horizon, I started back into the hull to take a rest before dinner. On my way down, I noticed Torsten, Malene, Henrik, and Kristian huddled in the cockpit. Malene was holding a device the approximate shape and form of a pneumatic bank tube: a clear plastic cylinder with a rubber hose running from its top out the bottom through two loosely attached covers at each end.

"What's that?" I asked.

"That's a Niskin bottle," Malene explained. "We're going to send it to the top of the mid-ocean layer, where we'll close it to get a sample of the water down there."

Although humanity had been aware of the Great Pacific Garbage Patch for nearly twenty years, this would be among the first of any research efforts to look for microplastic at such depth in the eastern North Pacific Gyre.[3] Could plastic be hiding here, unaccounted for by previous scientific research? I realized I would probably have to wait until the samples were processed at Kristian's lab in Denmark to find out.

After striking a deal with Rasmus that I would wash the dinner dishes solo if he could take care of our cooking duty, I grabbed my camera and notebook and joined the scientists. Kristian tied a long, slim rope to the Niskin bottle while Torsten began spooling the other end of the rope onto a winch on the bow. Each time about twenty meters of line slipped through her fingers, Malene slashed a red dash across it in thick permanent marker, and when she reached its tail, she secured the Niskin bottle with a tight knot.

Henrik and Malene stood back from the winch while Torsten slowly unspooled the rope. Kristian guided the Niskin bottle over the bow and into the waves until it disappeared beneath the cover of blue. When the unspooling rope revealed its two-hundred-meter red tick, Torsten held the winch and with his other hand clamped a small hinged brass weight around the line before releasing it to meet the Niskin bottle. "That should snap it shut," he grunted to no one in particular, pausing. When the line shivered, indicating the weight had made contact, he began hauling the device, now exponentially heavier with its seawater load, back up to the ship.

In the cockpit, Kristian laid an extremely fine piece of mesh over the mouth of a steel bucket, over which he carefully emptied the Niskin bottle. The water in the bottle looked fairly clear. If there was plastic inside, the particles would have to be nearly imperceptible in size—something to be examined more closely in the lab. When the bottle was empty, Kristian scraped the mesh and deposited the few visible pieces of suspected microplastic—and whatever unknown quantity of near-invisible nanoplastic that may have been caught—into an amber sample jar.

As the scientists worked, *Christianshavn* ebbed lazily toward Hawai'i like a languid piece of driftwood slowly weathering, in no hurry of reaching a shore. Without much wind available, her speed was rather slow—slow enough, in fact, to acquire a light smattering of barnacles and the green tarnish of algae, which crept up her hull. When *Christianshavn* crossed over the deepest part of the eastern North Pacific Gyre, nearly six miles of seawater separated the thick steel vessel from the seafloor. At the surface, the water was frequently glasslike, patterned by a nearly imperceptible ripple that reminded us we were dwelling in the middle of an unpredictable ocean.

Stillness and tropical sunshine combined rendered the hull unbearably hot during the daylight hours and the deck only slightly less uncomfortable before dark. As the late afternoon wore on, the sailors stripped down to their undergarments and dove, piercing through the skin of the water. We took turns to leap into the sea, though in such a calm it seemed there was little risk of the ship drifting out of reach. Two short-finned pilot whales, creatures born with the right equipment to navigate this realm, reeled effortless circles around *Christianshavn* and her human satellites with their agile fins and flukes. The whales' bulbous faces frequently broke the surface to eye the unusual creatures floating around them—us.

When the slightest breeze ruffled the sea, Torsten motioned for all the sailors to climb up the creaky plastic deck ladder back into the ship. *Christianshavn* would carry on arcing toward Honolulu, doldrums be damned. Her generously sized genoa sail could turn even the gentlest winds into a powerful propellant.

If we had chosen a different course, if instead the old ship continued sailing southwest past Hawai'i, out of the eastern North Pacific Gyre, past Guam, she'd eventually cross over a curving forty-three-mile-wide valley, a crescent moon etched into the seafloor. This is the Mariana Trench, the southern end of which is home to the deepest place on the planet, three strike above basins collectively known as Challenger

Deep.[4] Even there, nearly seven miles below the sea surface—in what is arguably the most remote place on Earth—there is plastic.

A few months before *Christianshavn* set forth across the eastern North Pacific Gyre, a hulking white-and-blue steel behemoth of a ship called *Tan Suo Yi Hao* cruised to Challenger Deep under the Chinese flag. There, her occupants, researchers from the Chinese Academy of Sciences, dropped an array of enormous Niskin bottles and seafloor corers, fixed to miles of thick steel cables, down to the deepest part of the ocean. The devices collected water and punched sediment before being pulled up from the deep sea to the deck. *Tan Suo Yi Hao* would return to Challenger Deep twice the following year to collect more seawater and sediment. The samples revealed that the depths contained levels of microplastic many times higher than those collected on the surface of the North Pacific.[5]

That same year, a different team of scientists, from Japan, published a black-and-white image of an intact plastic bag that had nearly reached the bottom of the Mariana Trench. It was discovered in a collection of photos snapped by a deep-sea remote vehicle in 1998 and reveals— more than six and a half miles below the sea surface—an indubitable emblem of humanity, in a place as far away from people as you can get.[6]

"There's a chance microplastic isn't as bad as we thought," Kristian reflected during our journey across the eastern North Pacific Gyre. "And then there's a chance that microplastic is much, much worse than we thought. We're working on getting to the bottom of this—now."

Some scientists on the quest to find the ocean's missing plastic—that is, the plastic unaccounted for by decades of surface sampling—have turned to computers for help. In 2016, a research group with partners in the Netherlands, UK, and US digitally simulated the movement and existence of plastic in the oceans from the moment it hits the surface, using the presently understood speeds at which plastics break up and how quickly they sink. Exposure to wind, waves, sunlight, extreme temperatures, plants and animals, and chemicals are all factors in the rate at which plastic fragments into microplastic and nanoplastic in the ocean. That year, the scientists estimated that 99.8 percent, or 196 million metric tons, of all the plastic believed to have entered the oceans since 1950 has settled below the surface, with 9.4 million metric tons sinking below the surface in 2016 alone. The continual fragmentation of plastic items into smaller and smaller pieces appears to cause rapid settling of plastic particles, particularly when shed from thinner plastic bags and films. These tend to break up more quickly than other types of plastic below the surface, the scientists found.[7]

If this modeling is accurate, most plastic that enters the ocean as floating debris sinks down rather quickly, snapping into smaller and smaller bits over time but never fully deteriorating. In fact, just-emerging research suggests there could be more than thirty times as much microplastic across the entire seafloor than is floating on the surface.[8] If we could, by some miracle of human will or invention, completely stop plastic from entering the ocean today, there would be none floating on the surface in two to three years. But below the surface, it appears that humanity's plastic legacy will continue to wreak havoc for an indefinite, possibly eternal, amount of time.[9]

The dynamic nature of the ocean presents plastic researchers with a perpetual challenge: The questions of how and when—or even if—they can use various kinds of scientific equipment to collect samples are most always dictated by an ever-shifting set of sea conditions.[10] Equipment is most accurate in stable sea conditions. If an ideal sea state presents itself, the seasoned scientist does not let it go to waste.

One evening while sitting beneath the starry sky on my watch, I could hear Malene, Torsten, Kristian, and Henrik inside the ship discussing whether or not to collect more water from the mid-ocean with the Niskin bottle. After brief deliberation, and consensus that the calm sea provided ideal research conditions, they decided to continue pulling samples into the night as the other sailors slipped into their bunks to sleep.

Such a nocturnal operation would require a lot more light than the stars and waxing crescent moon afforded. The ship's small solar panel hadn't yet charged up the ship's batteries enough to keep the powerful deck lights beaming for more than an hour. The sailors needed to power up the engine to recharge the ship's batteries, which would keep the lights running as long as necessary and also propel *Christianshavn* farther into the gyre. In the belly of the boat, the engine rumbled alive, reeking of diesel. Below deck, one could hear nothing but its roar, punctuated during its occasional sputters by footsteps thudding on the thick steel overhead.

My shift came to an end, and I headed into the noisy hull to my bunk. Earplugs helped sleep arrive, despite all the racket. But then sometime in the middle of the night, I was jarred awake by a loud string of Danish expletives shouted up on deck. "For helvede! For satan! Pis!" I tugged back my bunk's thick blue privacy curtains, perpetually misaligned on their flimsy plastic tracks, and rolled out to scramble up on deck. There, Torsten shooed me and several other concerned sailors back to bed with a pensive frown. Malene and Henrik loomed behind Torsten, peering over the stern and speaking in hushed tones. The rest of us—save for Rasmus, who, also frowning, rushed to join Malene and Henrik at the stern—headed back downstairs.

Early the next morning, Torsten roused everyone from their bunks, and soon the sailors were clamoring into the cockpit, pouring mugs of hot coffee. Malene and Peter were already awake, sitting together behind the wheel. In the darkness the night before, *Christianshavn* had entangled her propeller in one of her own ropes, Torsten announced solemnly, while the engine was switched on. Though we could manage to free the propeller, the incident had seriously damaged our engine, which he said was as good as gone, as our crew lacked equipment for making the necessary repairs. A few members of the crew traded anxious glances.

"What if some kind of emergency is to occur?" asked Sofie.

"How will this affect our research?" asked Kristian.

"Should we just turn around?" asked Chris.

The crew debated whether or not *Christianshavn* should turn back and sail on the winds, seeking safe harbor in Los Angeles—a briefer journey than continuing across the gyre. Torsten explained sea conditions were smoother heading west, and so, with the prospect of choppy seas on the return trip to the mainland US feeling more dangerous than pressing on into the peaceful gyre, the sailors ultimately agreed to continue forward. Though the idea of getting stranded in the Pacific Ocean on a ship without an engine did cross my mind, I trusted Torsten, a highly experienced sea captain, would be able to lead us through the ordeal.

*Christianshavn* and her inhabitants pressed on, carried by will and wind.

Despite relying on sail power alone to get through the sometimes near-windless gyre, two weeks after the prop-fouling, engine-busting incident, *Christianshavn* had made decent progress in her journey. And the scientists' sample jars, stashed in and beneath the leaky V-berth, were filling up one by one. By day nineteen, with dozens of scientific samples collected, the crew had less than one thousand nautical miles to travel before bumping up on the jagged shores of one of two central Hawaiian Islands, O'ahu or Moloka'i. With our crippled engine, we weren't sure exactly where *Christianshavn* would be able to make landfall. We left that decision to the elements.

I awoke in darkness around 5:30 a.m. on day twenty-two to what had grown to be the familiar tug of cool hands on my bare feet—my wakeup call to dress, clip on my self-inflating life vest, and step on deck. Though it was still about an hour to sunrise and the murky predawn obscured any fine details of the seascape, when I slid behind the wheel I could see—ever so faintly in the hazy purple horizon—the dark jagged outline of volcanic rock, the unmistakably solid silhouette of earthly terrain: land.

We'd made it to Hawai'i. . . .

Almost.

We'd still have to navigate, without an engine, precariously around the islands' sharp reefs and hidden rocky outcroppings certainly capable of ripping through our steel hull or beaching us, at best.

Rasmus glanced at the GPS monitor: There were two hundred miles of sea separating us from Honolulu. Hawai'i's capital city on the island of O'ahu, our original destination, possessed an abundance of marina slips and accessibility to shipyards where we could find the parts and services needed to fix our shot engine. But the jagged landmass toward which we were initially moving was Moloka'i—one of Hawai'i's least developed islands—which has a bumpy underwater topography and limited coastal infrastructure, making it less friendly to our injured sloop.

Our newfound proximity to land brought with it changeable weather and the remarkable presence of much other life. Blowing winds and churning waters full of fish conjured a chaotic cloud of birds all around: Boobies and tropicbirds rode the rising air currents, albatrosses whirled eddies in the yellow-pink sky, and as the sun rose higher, we could see endangered band-rumped storm petrels seesawing astride the frothy indigo sea. We'd intercepted these birds at a busy time, as they were apparently competing to fill up on local fish, squid, and plankton.

By the time I handed off the wheel to Chris around ten o'clock in the morning, the wind and waves had intensified, holding *Christianshavn* down in a heavy starboard list. All activities on board from that point on would be done with a right-sided lean. Torsten directed the crew to trim the sails and cut as close to the wind as possible to gain the speed needed to slip between the islands of Moloka'i and O'ahu. If we could do that, there was a small but existent opportunity to reach a more desirable harbor in Honolulu.

Sometime around midday, a handful of sailors gathered in the cramped, wood-paneled saloon, celebrating an impending arrival in Hawai'i. Suddenly, we heard a deep *CRACK!* as everything and everyone on the boat was thrown starboard. Through the hatch, I saw Chris at the helm, frantically turning the wheel hand over hand, his face screwed into a knot of bewilderment. His hands turned the wheel uselessly on its axle. The crew rapidly let out the sails, bringing *Christianshavn* to an abrupt halt. Torsten dropped into the hull from the deck, racing to the equipment room to examine the inner workings of the rudder, which appeared to have lost contact with the wheel. He crawled out and called in Rasmus, and then Peter, to take a look. Each remained silent while inspecting the rudder in the dim, damp, and cramped space.

After ten minutes, Peter unveiled to the crew a jagged, peg-like chunk of metal wrapped in a greasy rag, which he proffered like a twisted oddity, something both miserable and fascinating to look at. It was a piece of the rudder, which had been jammed out of place, several of its parts broken and bent by age and the sheer power of the sea. Rasmus said it was possible to loosely steer by gently moving the wheel within a narrow set of degrees.

But now if we pushed the ship too close to the wind, or were knocked around by a big wave, the rudder could completely lose contact with the wheel. If that happened, we'd be set precariously adrift—liable to crash into Hawai'i's abundant rocks and reefs.

"Erica?" I heard Kristian softly calling me over the din of Lou Reed's deadpan crooning in my ears and roused myself from slumber. I'd often listen to music while sleeping to drown out Torsten's snoring in the bunk above me. I checked the time. I'd been asleep a few hours.

"Hey, sorry to wake you, but you need to know this," he frowned and paused. "The rudder is completely gone. We're generally heading to Honolulu. We're going to see if we can adjust the sails and make a makeshift rudder."

"So, we're getting there?" I winked, trying not to show that what had been a tiny pit of worry had begun to sprout a tree of anxiety in my stomach.

"You could say that," he said with a weak smile. "Torsten asks that all hands get on deck so we can hopefully get there without any problems."

I pulled on my clothes, still damp from my most previous shift, as I climbed on deck. They felt cool against my skin despite the tropical heat. The wind had picked up significantly; the seas grew rough. Some of the swells surged at least twenty feet into the air. Rasmus sat on the stern bench holding a long wooden plank, which was fastened to a larger piece of wood behind the ship. He had pulled several spare wooden boards from his stockpile below the stern, which he'd nailed together to create a hand-controlled rudder—a "Viking rudder," he called it. The wooden contraption seemed to be moving *Christianshavn* generally in the direction she needed to go. She was *moving*. A glint of relief shone in the sailors' weary eyes.

While all hands made their way to the deck, Torsten stood in the navigation room, hunched over the chart table searching frequencies on the ship's radio. He held the device to his ear and listened briefly, before moving it to his mouth. "US Coast Guard, this is *SY Christianshavn*: We have no engine, no rudder, over."

Torsten flipped off all the ship's power, save for that charging his gear in the navigation room. It was starting to feel like an emergency—and that maybe a rescue was merited. Torsten remained silent when he joined the rest of the crew on deck.

In what seemed like mere moments, a white steel vessel—its bow adorned with a thick slash of red paint before a lesser slash of blue—appeared on the horizon, half-concealed in the enormous blue swell: Coast Guard cutter *Kiska*.

Torsten noticed all eyes on *Kiska*. "We are not being rescued," he told the crew plainly. "They are just watching over us and warning other vessels in the area that our steering is impaired. I see no need for us to abandon ship. We will call a tow when we get to the calm waters of the harbor."

The crew nodded in agreement and continued lashing the sails, ropes, and equipment to the deck to prepare for landfall. The large rolling swells undulated like steep hills, and *Christianshavn* braved them headfirst with her strong steel hull. Suddenly, the force of a rogue wave cracked the wooden Viking rudder into a few chunks of splintered wood.

And then we had nothing.

Torsten crawled into the saloon and scoured shelves of stained and torn books—mostly written in Danish—that comprised the ship's library. He pulled a tattered blue tome from the shelf and leafed through it with one hand—the other he was using to secure himself as he sat wedged between the saloon table and bench—as heavy seas tossed the helpless ship. Finally, Torsten announced, he had found a possible solution: Drag the spinnaker pole behind the ship with plenty of heavy lines. The extra weight could maneuver us in the right direction.

The new plan from the old book succeeded, and *Christianshavn* pressed on. Torsten again called all hands on deck as we continued toward land. Heavy sails needed to be yanked down, folded up, and packed away; ropes, coiled and stowed, all while the ship bucked wildly over rough seas. Clearing the deck in such conditions felt nearly impossible.

I struggled to scrunch up the cumbersome, waterlogged jib into a tight bundle and secure it to the railing, maneuvering on my hands and knees, as the ship tipped into the water so steeply that at times I felt like I was clinging to a slippery mountainside. Malene noticed me fumbling with the jib and crawled over to help. Each time a gigantic swell took aim at *Christianshavn*, she would shout "bølge!" ("wave!"), a warning to the other sailors to brace themselves.

At one point I noticed something mobile and gray not too far in the distance—it rose up out of the water and pressed its shining body against the cloudy sky for just an instant: a dolphin. It turned out to be one bottlenose dolphin, part of a pod of thirty or so individuals. They were on the move, surging up above the swells in neat leaps or doing elegant midair twists and body slams, giving a little tail shimmy just before hitting the water with a splash.

We were not alone. Filled with a surge of energy, Malene and I finished folding up the sail and tied it tightly to the bow's railing, before moving inside the hull to wait out the very last hour of our journey to Honolulu. With the help of a towboat, *Christianshavn* finally limped into Kewalo Basin Harbor.

After our ship was dragged to a temporary concrete slip and briefly inspected by the Coast Guard, we rushed off the deck and scrambled onto the pier to move and stretch our legs, no longer restricted to a space just fifty-four feet in length, for the first time in twenty-three days. The ground turned still below our feet, though our bodies, so accustomed to life at sea, attempted to keep moving, which sent us swaying like reeds in a breeze.

As the Pacific Ocean and its creatures revealed, there is a plastic crisis at sea. Having completed our journey of more than three thousand nautical miles across the gyre, back on land we could probe the problem more deeply. Still necessary was an investigation into plastic's full range of risks, and efforts to track exactly where it was coming from. Plastic is not made in the ocean, after all.

Immediately, the slow life we had known on the water clashed with the fast-paced life we had just returned to on land. Instead of just eight others, there was an unending stream of people floating around our periphery; instead of the melodies sung by waves, wind, seabirds, and the clinking of sail shackles, there was a general buzz, an urban din; instead of the simplicity of a deck, hull, the endless sea, there was a maddening blur of roads, skyscrapers, supermarkets, malls, beaches, restaurants, bars; instead of austerity and adaptation, there was luxury and waste. Back on land, I missed the sea.

Dr. Sherri A. "Sam" Mason, professor of chemistry and leading researcher in freshwater plastic pollution, inspects suspected particles of microplastic pulled from sediment samples from the Great Lakes at State University of New York Fredonia in June 2017. Photo by Erica Cirino.

# PART II

# Little Poison Pills

CHAPTER 5

# Pick Up the Pieces

Science labs are tidy, stark, light, and white walled, occupied by punctilious, white-coated people. Sailboats, especially old ships like *Christianshavn*, are perpetually knocked topsy-turvy by waves, slickened by saltwater, dark and wood-paneled, inhabited by less-refined—often wayward—types, the seafarers among us.

In Kristian Syberg's lab at Roskilde University in Denmark, these realms mixed before my eyes: A spread of petri dishes was lined up across the lab bench. Each of the glass circles held piles of familiar microplastic pieces, pieces *Christianshavn*'s crew had pulled from the Pacific Ocean two years earlier. Small artifacts from the sea had entered the lab. Torsten, the ship's scruffy captain—now fully bearded and long haired—stood buttoning up an ill-fitting white lab coat. A sailor had pirated the lab.

It was a dreary winter day in Denmark, the usual Scandinavian mix of cool and wet and gray. And though the surroundings were now starkly different from those aboard *Christianshavn*, it was impossible to pore over the small beads and fibers and fragments and films atop the lab bench without ruminating on weeks spent exploring the vast Pacific Ocean. All these little plastic bits seemed familiar, somehow, and a few pieces were indeed recognizable—the fraying tangle of clear monofilament, the cobalt-blue microbead, and the star-shaped dusty-pink shard. Under Kristian's watchful expert eyes, Torsten sailed into the throes of some serious scientific déjà vu.

While at sea it seemed like pulling microplastic from the waves was a slow process, studying those particles in the lab isn't much faster—even though there are no nets to cast or slippery decks to navigate. First of all, there were many samples to work through: On the North Pacific Gyre expedition, the crew had thrown the manta and vertical trawls in the water nearly every day, several times a day, unless the water was too rough or too calm for their trawling equipment, which required the ship to be moving at a slow clip of two to four knots to work properly. Then there were several samples of plastic pulled when the sailors had dropped the Niskin bottle.

Torsten also planned to analyze additional samples collected by a crew that he, Malene, Rasmus, and I joined on *Christianshavn* again in 2017. After extensive repairs and a full repainting, the old ship had continued on her way from Hawai'i into the South Pacific, over the equator to Nuku Hiva, one of twelve small, remote volcanic islands that make up the Marquesas. This archipelago rises out of the Pacific about nine hundred miles northeast of Tahiti.[1] Scientists studying plastic pollution in that region of Oceania think it's likely that the North Equatorial Counter Current, which wraps west to east around the globe just above the equator, carries a massive amount of plastic debris from the highly polluted, highly populated region encompassing Indonesia, the Philippines, China, Thailand, and Vietnam and farther east into the relatively more pristine and less populated islands of Micronesia and Polynesia. This plastic is carried far and wide, including to Hawai'i and elsewhere in the eastern North Pacific Gyre, which we previously explored.

Further complicating matters in the lab, we had to clean each of the algae-coated particles with an alkaline solution (which a lab tech had warned us to keep away from our skin and eyes). This would help reveal whether each particle was actually plastic, and not a speck of wood, shell, or seaweed—which are also prevalent in the seas but would dissolve a bit in the solution. Like so much contemporary microplastic research, this would be an experiment done with most of the human senses, potent laboratory chemicals, and fine-tuned scientific equipment. "Vertical trawl, 11/9, tube 7," Torsten announced, holding the jar up to the light. Several colorful confetti-like particles—quite possibly microplastic—swirled around inside. In the spirit of at-sea camaraderie, I lent a hand, scribbling his words onto the tally sheet used to keep track.

Sitting at the lab bench, the captain emptied the sample into a stack of mesh sieves, like those we used at sea. One by one, he picked up tiny fragments, foam bits, and film pieces from the mesh and peered at them under a dissecting microscope. Torsten checked for boxy shapes—a sure sign of a plant's energy-making chloroplasts—and more rounded, repeating cells, which characteristically compose much of animals' body tissues. When he found a short fiber, he flicked a lighter beneath one end of a steel needle and touched the hot metal to the pinkie-nail length thread to see if it would melt (indicating plastic), or burn or remain unchanged (indicating something else). Torsten squeezed all the fragments with a steel pincer to see if they'd easily shatter, to rule out pieces of shell and boat paint.

After this first round of scrutiny, he placed, one at a time, each suspected piece of microplastic into the Fourier-transform infrared (FTIR) spectrometer, a device used to reveal the primary chemical composition of materials. FTIR spectrometers have been used since the mid-1960s, primarily by scientists hired to test industrially manufactured materials for consistency. Yet over the past decade, as plastic pollution research has rapidly ramped up, FTIR has gained a foothold as scientists' go-to tool for identifying the provenance of seemingly unidentifiable plastic particles. The spectrometer, itself encased in plastic, was about the size of a large printer and contained mirrors, a crystal, and an infrared beam. The beam bombards a scientist's sample material—in this case, a suspected microplastic particle, which is placed on a platform that presses it to the crystal. When hit by the light beam, the sample material absorbs infrared energy of specific wavelengths and begins to vibrate. A computer reads the vibrations in less than a minute and translates them into a visual picture of peaks and troughs, which scientists are trained to decipher. The waves carry chemical clues revealing from which items— clothing, bags, bottles, containers, fishing gear—microplastic particles have shed.[2]

The press of a large green button on top of the FTIR spectrometer commenced the release of infrared light, which the sample absorbed, sending a digital wavelength zigzagging across a computer screen. Finally, here were some answers.

"Blue fragment: high-density polyethylene; clear film: low-density polyethylene; green fiber: nylon; white fragment: polyvinyl chloride. . . ." As Torsten called out his results, a dismal mosaic began to assemble itself from the weathered plastic pieces sitting in the petri dishes on the lab bench. The pattern depicted a small snapshot of a vast ocean of microplastic, a fraction of the trillions of particles broken off from the ubiquitous synthetic items with which humans fill their lives: bags, containers, packaging, electronics, household items, children's toys, fishing line, building materials, clothing, lifesaving medical equipment, and useless knickknacks alike.

Kristian peeked at a few fragments of high-density polyethylene, the variety of plastic typically used to make hard plastic containers, under the microscope. The jagged, pitted surfaces provided lots of area on which the plastic was likely carrying toxins, he noted to Torsten.

Plastic debris of all sizes can physically block movement of food through animals' digestive systems, causing starvation over time.[3] All of us who had sailed aboard *Christianshavn* were familiar with Chris Jordan's photographs of sliced-open albatross corpses—laced with bottle caps, lighters, utensils, indeed, all manner of plastic stuff. But wild animals' consumption of microplastic is particularly concerning because of the small particles' tendency to pass chemicals used in plastic manufacturing and acquired in nature into the bodies of the living beings who consume them.[4]

Many plastic-manufacturing chemicals, additives including plasticizers, are known as toxic, as proven by studies on nonhuman animals and people alike. Bisphenols, like bisphenol A (BPA), and phthalates are two common classes of plasticizers known to interfere with hormone activity in wild and laboratory animals, leading to metabolic and growth problems, as well as cancer.[5]

Those chemicals commonly adhering to microplastic particles in the ocean include pesticides, such as dichlorodiphenyltrichloroethane (DDT), which has been banned worldwide except in cases of controlling insect-borne epidemic diseases like malaria. Other toxins that adhere to microplastic include industrial chemicals like polychlorinated biphenyls (PCBs), which were used for myriad purposes, including additives to anti-fouling ship-bottom paint until most nations around the world—including much of the European Union, and the US—prohibited most of its uses. Both DDT and PCBs are toxins known to cause adverse health effects in people and wildlife through direct exposure routes such as inhalation and ingestion. These and many other manmade toxins now found in nature are classified as "persistent organic pollutants" (or POPs), possessing a chemical structure resisting degradation, allowing for long-term retention and circulation in the environment. And so POPs have the ability to poison continuously, penetrating various and multiple levels of the living world before being converted into related chemical byproducts (called metabolites, which are usually also toxic).[6]

When a living creature eats plastic, scientists believe that the toxins manufactured into or absorbed by plastic in nature leach out into their bodily systems, possibly causing health problems.[7] For example, marine creatures who consume microplastic are also dosed by the particles' plasticizer ingredients and any POP hitchhikers from the sea. Compared with more intact plastic items, microplastic is lent additional surface area in its jagged edges that may enhance its chemical-delivering effects into animals' bodies.[8]

This phenomenon, which Kristian has studied, is called the *vector effect*. Molecules of these toxins, carried by an itinerant vector—plastic—are often lipophilic, or have a physical and chemical propensity for fats, and thus tend to accumulate in living beings' fatty tissues. Fat acts as a repository for toxins—one unlocked when the body taps fat cells for energy. When that happens, the liver must metabolize the chemicals trapped inside.[9]

Food scarcity is a common experience for wild creatures dwelling in and above the vast and unforgiving open ocean. While working with terrestrial wildlife, I've witnessed animals like hawks and owls quickly fall ill or die during lean periods, though not of starvation per se: Toxicology reports run on predator animals have revealed that poisons in their bodies' fat cells, accumulated over a lifetime of eating wild food, become unlocked when their body burns fat for energy. Because they don't degrade quickly, POPs linger and accumulate in the body over time. When an animal must live mainly off fat reserves, the rate at which the body metabolizes toxins speeds up, sending harmful chemicals once stowed away in fat cells to the brain and other vital organs.[10]

Of course, the Earth is made of chemicals. *We* are made of chemicals. It's the toxic chemicals surrounding us that we must avoid to stay safe. Some of these, like radioactive metals, occur naturally in varying amounts, while others, like PCBs, are introduced to the world by people. These chemicals are now around us, and inside us, at all times, and we can be harmed if we're exposed to too much at once or a great enough concentration over a lengthy increment of time. Microplastic's ubiquity and chemical-magnifying vector effect complicate this equation, filling our world with small doses of poisons that—if absorbed by our bodies—push us closer to the thin knife-edge that separates safety from harm.

A number of scientists are presently studying the role of the vector effect in the open ocean, exploring how plastic is spreading poisons throughout the sea and marine animals' bodies. Ecologist Chelsea Rochman, preeminent plastic researcher at the University of Toronto, is one of them. I once asked her which chemicals commonly detected in microplastic are most concerning. She answered quickly: "Plasticizers. Many scientists are focused on other chemicals that microplastic may absorb once in nature, but we also and maybe more pressingly need to take a hard look at what kinds of plastic additives are now being introduced to the environment through plastic, especially microplastic and nanoplastic."[11]

The vector effect is challenging to study for a number of reasons, including the great variety of toxic chemicals potentially carried by a piece of microplastic. This research is also trickiest when it involves the very smallest pieces of plastic—nanoplastic—which are challenging to collect and analyze. Scientists have tracked nanoplastic particles consumed by plankton that are then consumed by fish. And in the fish's bodies, scientists have tracked the nanoplastic as it moves from the digested plankton in their guts to their bloodstreams and then to their brains. Fish with nanoplastic in their brains display aberrant behaviors such as a reduced or sped-up feeding time, and too much or too little exploration of their surroundings. These behaviors, which often waste energy and impair a fish's ability to hunt for prey, seem indicative of neurological malfunctioning—likely caused by the presence of plastic and the chemicals it carries.[12]

Nanoplastic particles don't necessarily have to be eaten to cause harm. They have been absorbed by fetal fish still growing inside their eggs. Scientists at Duke University have observed that zebrafish eggs plunked into nanoplastic-spiked water just six hours after being fertilized transferred some of the nanoplastic from the surrounding water into fish's embryos. There the nanoplastic migrated throughout the still-developing fetuses. When the baby zebrafish hatched, they appeared physically normal. But below their skin, their hearts were beating abnormally slow, and, unlike healthy young zebrafish—which are usually energetic— these young fish moved rather languidly around their tanks. These characteristics would undoubtedly reduce their chances of survival if they were trying to hack lives in the wild.[13]

While the hazards of microplastic and nanoplastic ingestion and exposure are still not completely understood, scientists agree these small plastic particles are efficient transporters of toxic chemicals that pose a threat to the lives of wild animals.

Once the last trawl sample had been analyzed, Torsten covered and stacked the glass petri dishes for safekeeping. Torsten had counted and analyzed hundreds of pieces of microplastic—which seemed at once an enormous amount and just a drop in the ocean. It left me wondering: If it was possible to collect a hoard of plastic particles in one of the most notoriously polluted regions of the seas over the course of just a few weeks, how much plastic could be found circulating Earth's freshwater ecosystems—and how might those waters be affected by its presence?

# CHAPTER 6
# Troubled Waters

From the shore, the rippling blue surface of Lake Erie seemed to stretch endlessly into a gray, hazy horizon. For a moment, I forgot that my feet were planted on solid ground, and the water that unfurled before me seemed as vast as the sea.

The air was charged electric, a thunderstorm and associated downpour looming. A loud, not-too-distant boom finally persuaded me to slip back into my rental car and get going. I'd stopped by the lake's edge on my way from Buffalo, driving south to meet Dr. Sherri A. "Sam" Mason, who was then a chemistry professor at the State University of New York (SUNY) at Fredonia. (Later, she'd work as sustainability coordinator at Penn State Behrend.) Recently, I had learned about Mason's plastic pollution research that was quickly revealing a massive pathway of plastic into the ocean that had long been ignored: Earth's freshwater bodies.

I made it into the refuge of Mason's big-windowed lab just before the skies opened up and a downpour saturated the scorched summer soil. Amber jars, which according to their labels held sediment dug out from the bed of Lake Erie by the US Geological Survey, were lined up on her bench. Mason, who'd whisked on a white lab coat before entering the room, tied back her long auburn hair and unscrewed the first sample, releasing the fetid stench of low tide.

"We've got to separate the solids from the liquids, then dry the solids, then look for microplastic, then closely examine what we find to verify that it's actually microplastic—and determine what kind of plastic it's made of," she said, tipping the muddy contents of the jar out into a fine mesh sieve. She dripped freshwater into the jar with a dropper bottle to catch every last bit of what was inside, then swished and emptied it. Just like in the open ocean, scientists have studied the amounts of plastic present on lake surfaces for decades. At the time of my visit, experts like Mason were just beginning to look for plastic in freshwater sediments—a place where perhaps much plastic was hidden out of sight. "More plastic found in freshwater sediments would mean a greater overall plastic load, and thus a greater effect on aquatic ecosystems— both saltwater and freshwater," she said.

Most large lakes empty into rivers or streams that lead to the sea. Plastic follows this route.[1]

"Take microbeads, for example," Mason continued, extracting a glass jar filled with what looked like tiny neon-pink sugar sprinkles from one of her drawers. "These are microbeads. A microbead is just a small, rounded piece of plastic. When microbeads are rubbed against skin or teeth, they abrade. That's why you'll commonly find microbeads in toothpastes, body scrubs, and other cosmetics." Microbeads contain plasticizer chemicals and other additives like colorants, as do all items made of plastic, she said.

Basically, Mason told me, when you use toothpaste, a skin scrub, or other products that contain microbeads (usually indicated on the label as some variant of *polyethylene* or *polypropylene*), "you're smearing plastic and the chemicals it contains all over your teeth or skin. Then you rinse and wash the microbeads down the drain."

Because the majority of wastewater treatment plants weren't designed to capture microbeads and other microplastics, the tiny plastic particles escape with treated effluent into the environment. In the Great Lakes region, wastewater treatment plants flush their effluent straight into those gargantuan freshwater bodies they have at the ready: Lakes Michigan, Huron, Superior, Ontario, and Erie, which flow into the Atlantic Ocean via the St. Lawrence River. A research team led by Chelsea Rochman at the University of Toronto has estimated that American wastewater treatment plants flush around eight trillion microbeads out with effluent, directly into aquatic ecosystems—creeks, streams, ponds, rivers, marshes, estuaries, lakes, and the oceans—every day. Rochman and her team also noted that microbeads are highly concentrated in the sludge—or solids—collected by wastewater treatment plants. In the US, and many other parts of the world, this sludge is often repurposed as agricultural fertilizer, and when it rains, hundreds of trillions more microbeads are expected to wash from land into Earth's aquatic ecosystems.[2]

Salts, muds, waxes, sugars, clays, and ground-up shells or nut husks can produce the same desired exfoliating effects as microbeads. But plastic microbeads are usually a far less expensive choice for health and beauty companies, which began formulating products containing the tiny plastic spheres in the 1980s.[3] About two decades later, scientists began emphasizing that microbeads were a ready-made source of plastic pollution about the same size and shape as plankton, a major food source for many aquatic animals.[4]

I watched Mason weigh, inspect, and label several of the mud samples for drying in a countertop oven, the next step in her research. "I do expect to find lots of microbeads," she said. "Also microfibers, which similarly wash out with wastewater into the Great Lakes. Every time you wash clothing made with plastic fibers—think, a polyester fleece jacket or spandex leggings—a significant number of plastic fibers shed off and drain out of your machine along with the washing water."

According to some scientists, a typical load of laundry—about thirteen pounds (or six kilograms) of mostly synthetic textiles—could shed more than seven hundred thousand plastic fibers. Most modern washing machines aren't designed to capture plastic microfibers, which look like tiny threads, and particles of microplastic, like glitter.[5] What's more, when a piece of clothing gets worn out, is deemed out of fashion, or no longer fits, it is most often tossed in the trash. Several concerning issues have long plagued the fashion industry: slave labor, dangerous working conditions, and toxic dyes among them. We can add plastic waste to the list.[6]

For much of the 1900s and early 2000s, cotton dominated US textile imports. But by 2014, synthetic textiles—made of polyester, acrylic, nylon, and similar man-made materials—had surpassed cotton as the country's most-imported textile materials. Manufactured primarily in Asia, synthetic textiles are less expensive than cotton.[7] According to the US Environmental Protection Agency (EPA), clothing accounted for up to 6.3 percent of all municipal solid waste, about 16.9 million tons, produced across the nation by 2017. That year, more than 11 million tons of clothing were landfilled, 3.2 million tons were incinerated, and just 2.6 million tons were recycled.[8]

Beauty and fashion, it seems, will be major contributors to humanity's plastic crisis for some time to come. Mason and a few collaborators would later determine plastic fibers, shed from clothing and also fishing gear, to be most prevalent among all microplastic particles found in the sediments of Lake Erie and Lake Michigan.[9]

In addition to uncovering the extent to which microfibers and other types of microplastic lay in the lakes' sediments, Mason said it was important to test microplastic particles found in the Great Lakes, particularly Lake Erie—which is surrounded by industrial complexes on its US shores—for toxic chemicals.

While the US Clean Water Act of 1972 tightened regulations on industrial discharges and pollution into America's surface waters, and freshwater quality has generally improved since then, Lake Erie and other water bodies are by no means pristine.[10] Like the oceans, freshwater bodies are contaminated with all manner of industrial pollutants known to harm environmental and human health. Water samples scooped from various regions of the Great Lakes reveal the presence of dozens of concerning long-lasting chemicals, as well as pharmaceuticals like antidepressants and antibiotics.[11]

At the same time, these polluted waters play a critical role in supporting life. An estimated eleven million people drink water pumped from Lake Erie.[12] More than 3,500 species of plants, animals, and fish live in and around the Great Lakes.[13]

Near industrial areas in the Great Lakes region, pollution routinely rains down onto the soil, saturates the air, and leaches into waterways—where it appears to adhere to plastic. On the surfaces of microplastic particles scooped from Lake Erie, Mason and a collaborator, Lorena Rios Mendoza of the University of Wisconsin–Superior, have detected polycyclic aromatic hydrocarbons (PAHs)—a class of chemicals emitted during combustion—as well as now-banned PCBs and DDT.[14]

"Little poison pills," Mason called the microplastic particles.

When I met Mason, the extent to which these "pills" were affecting local wildlife—and people—was yet unknown. In 2012, she'd trawled the surface waters of the Great Lakes and discovered concentrations of microplastic higher than those present in some areas of the notorious Great Pacific Garbage Patch—more than 230,000 plastic particles per square kilometer in Lake Ontario, and some 45,000 particles per square kilometer in Lake Erie.[15] Studying pollution in the open ocean is critical to understanding more about how plastic moves and exists around the globe—but we also can't ignore freshwater bodies like the Great Lakes, upon which so many lives, human and nonhuman, depend.

Mason was born in Texas and lived in Montana for a stretch, but she's grown close to the Great Lakes over the years, particularly Lake Erie.[16] That very morning, she had risen before sunup to squeeze in an early swimming session. Mason had decided that this particular summer was a good time to swim fifteen and a half miles across Lake Chautauqua and over twelve miles across Lake Erie to highlight freshwater ecosystems as major accumulation zones for microplastic, which her research continues to reveal.

Her swims were also political statements; shortly after Donald Trump's inauguration to presidential office in January 2017, he quickly moved to reverse the nation's progress in protecting and remediating the environment by defunding the federal programs and agencies tasked with caring for it. In what would quickly become an infamous and incessant string of deep budget cuts to environmental research and protection, Trump moved to eliminate funding for a major federal water conservation program and cut the EPA's budget for clearing pollution from the Great Lakes by 97 percent.[17] Mason reiterated that people have come a long way in cleaning up the Great Lakes, and we can't afford to reverse progress now.

Over the following months, Mason's experiment would reveal that Great Lakes sediment contains microplastic as ubiquitously as the surface of the lakes. She found thousands of plastic bits, primarily microfibers and high-density polyethylene fragments, in the jars of lakebed mud, as confirmed by analysis with an FTIR spectrometer, like that I'd seen in Kristian's lab. The existence of so much microplastic in Great Lakes sediment, Mason explained, is problematic for a variety of reasons.

"Lakebeds may seem like quiet and murky places without much going on, but really they're teeming with life that keeps a lake's food web going. Even if microplastic particles are embedded in sediment, creatures that dwell in the depths like bottom-feeding fish, worms, and mussels will inevitably be exposed and are likely consuming it," Mason explained. "We've long been underestimating the scale and implications of microplastic pollution, especially here in the Great Lakes."

In Lake Erie, one creature that seems to regularly encounter micro-plastic is the daphnia. Sometimes people refer to them as water fleas because they're somewhat similar in size and body shape to dog fleas; however, these aquatic zooplankton are not parasitic.

Though they're only the size of a sesame seed, daphnia—which can belong to any one of hundreds of species—have enormous appetites. Life for daphnia primarily consists of breeding, zigzagging through surface waters by flicking their branched antennae, and voraciously consuming algae, yeasts, and bacteria—which would bloom large and unbridled, depleting waters of oxygen and sometimes causing disease, if left unchecked. Females lay clutches of eggs every three to four days over the course of their lives, which last from just days to months. At some point during their brief lifetimes, many daphnia are eaten by larger zooplankton, amphibians, insects, and small fish, which are in turn eaten by birds, and larger fish—which are eaten by us.[18]

Just like in the ocean's middle layers and seafloor, increasing quantities of microplastic appear to be accumulating in the Great Lakes' waters and sediments. It seems that freshwater zooplankton like daphnia and other aquatic creatures may be unable to avoid it. But just because daphnia encounter microplastic, does that mean they're also eating it? While Sam Mason examined sediments from Lake Erie, graduate biology student Heather Barrett investigated this question at SUNY Fredonia. Barrett spent days hunched over a lab sink alternately filling, swishing, and draining vials of water mixed with toothpaste, body wash, and soap, in an attempt to separate suds from substance. She was seeking out plastic microbeads to feed to her brood of daphnia, creatures so small they are easily killed by even a drop of the highly concentrated cleansing chemicals that these products contain.[19] Hence the need to vigorously rinse the microbeads found in soaps, pastes, and scrubs before offering them to the tiny creatures.

"The whole lab smelled like Bath and Body Works' Country Apple soap for weeks," Courtney Wigdahl-Perry, Barrett's graduate advisor and experimental collaborator at SUNY Fredonia, recalled with a laugh.

When the products continued foaming after multiple washes, Barrett relented and instead opted for plain manufactured neon-yellow microbeads, sans soap, which would also be easy to track should the clear-bodied daphnia consume them. Manufactured microbeads vary in size, and those Barrett and Wigdahl-Perry used were exceptionally small, each about half the size of a small grain of table salt; too minuscule to see individually with the naked eye. But that was just the right size for daphnia to eat. The team opted against feeding the daphnia microbeads pulled from the Great Lakes, Barrett said, because they wanted to see if daphnia were compelled to eat plastic independent of whether or not it was coated with the tantalizing DMS coating of algae known to compel other animals to eat microplastic.[20]

Decades of research have demonstrated that different species of zooplankton, including daphnia, can't choose what their branchlike feeding appendages snag as they drift through water. But they do possess the wherewithal to use their small claws to select which parts of their catch they want to eat—discriminating based on size and perhaps even taste—and which they want to discard. They toss those select morsels small enough to eat—bacteria, yeast, or algae—into their mouths. When daphnia's preferred food sources have colonized a small enough piece of microplastic, however, daphnia's task of sussing out what is nutritious food and what is something that could harm them grows more complicated. Brand-new microbeads would put daphnia's food-selection skills to the test and possibly help researchers understand: Do daphnia eat microbeads because of the plastic's algae coating, or are they attracted to microbeads themselves—or both, or neither?

Barrett and Wigdahl-Perry lined up several small glass bottles of freshwater on the lab bench. Each held daphnia—which immediately crowded together at the surface. To each bottle, the scientists added a dash of green lab-grown algae, to encourage the daphnia to eat, and in a few of the bottles they sprinkled generous helpings of the neon-yellow microbeads. The microbeads floated at the surface around the daphnia while the algae sank. But the daphnia weren't eating. That was no surprise, Barrett said, as daphnia need a gentle water current, or must create such a current with their appendages when there isn't one, in order to catch food. Wigdahl-Perry and Barrett slid the bottles into a metal contraption called a grazing wheel, a scientific device resembling something like a miniature Ferris wheel for plankton placed in lab samples, and started its motor. Round and round the bottles went—better distributing the daphnia, algae, and microbeads in a current—which quickly encouraged the daphnia to eat as they would naturally in the more dynamic waters of a lake or ocean.

Within minutes, the guts of daphnia suspended in the bottles containing plastic turned bright yellow. Some of the daphnia were absolutely packed with microbeads. Now they were eating. Yet in the bottles with added microbeads, much algae went untouched, suggesting those daphnia that had consumed microplastic were experiencing a diminished ability—or perhaps preference—to eat real food. Their counterparts in the bottles lacking microbeads contentedly munched through the nutritious algae they were offered. If continually fed microbeads, it seemed these daphnia would likely continue to prefer it over algae, and the plastic would not provide them with enough energy to reproduce, let alone survive, beyond a short stretch of time.

The daphnia that were fed microbeads, unlike the daphnia solely fed algae, were vulnerable to physical blockages of their digestive tracts. These plastic obstructions emerged as visible yellow bulges in several of the daphnia's guts, ceasing all digestive movement—a major injury and possibly lethal consequence of consuming microbeads for an animal this small. While the waters holding the laboratory daphnia were inundated with quantities of microplastic more copious than their wild cousins encounter in nature, even just one single piece of microplastic can spell disaster. When examining a net full of wild daphnia pulled from Lake Erie, Barrett and Wigdahl-Perry discovered one individual that had its back claw tangled in a single microplastic fiber, which could be enough to prevent it from feeding—leading to starvation.

On a more hopeful note, Barrett said that when she pulled the daphnia out of the microplastic-spiked tanks of water and moved them into a clean tank, nine out of ten cleared the microbeads out of their bodies within a few days—and even went on to reproduce. Yet her research only examined the physical effects of microplastic exposure to daphnia. She could not be certain whether toxic plasticizers or other chemicals that may have attached themselves to the microbeads remained and could possibly cause problems, even after daphnia excreted the plastic.

"Clearly, daphnia are eating microbeads and can be physically harmed if constantly exposed," said Wigdahl-Perry. "There is probably a lot more going on here. This is a starting point, and now we should take steps to look at the process happening in the real world."

If daphnia are indeed attracted to eating microbeads—which are extraordinarily abundant in the Great Lakes—as Heather Barrett and Courtney Wigdahl-Perry observed, what kind of chemicals were zooplankton exposed to when they consumed them? Additional information was needed to nudge the scientists closer to understanding what was happening in the real world, to wild daphnia and other links in the Great Lakes food web—and all other freshwater environments—that have come to be filled, top to bottom, with plastic.

As the pair watched the hungry daphnia gobble down microbeads, an hour's drive north at the University at Buffalo, Joseph Gardella Jr., a prominent chemistry expert who has worked at Buffalo since 1982, and chemistry PhD candidate Abigail Snyder were screening microbeads for chemical hitchhikers. With state-of-the-art scientific instruments usually reserved for use by materials scientists and industries, Snyder examined the porous surfaces of microbeads scooped from the Great Lakes by Sam Mason, the essential link who brought these two projects—one focused on biology and the other on chemistry—together.

Mason, a top chemist herself, is also one of the world's foremost experts in freshwater plastic pollution. Her research occurs in the ill-defined borderlines between the two scientific disciplines, biology and chemistry, in the place where plastic pollution and many other man-made problems exist. With several years of funding from New York Sea Grant, Mason united the two schools of research, and two pools of researchers. Her position is one that, consciously or not, further emphasizes the fact that cross-field scientific cooperation is the only way to get a clear picture of how the workings of an already-complex natural world are affected by the even more complicated and unnatural activities of humans.

At the University at Buffalo, Snyder used several gargantuan machines to test for chemicals on the surfaces of the tiny microbeads. Each metal contraption stood several feet tall on top of the lab bench and was outfitted with an array of wires and cylinders, arms and gauges, tiny screws and enormous bolts most usually typified to belong under the purview of a mad scientist.

These complex instruments, one performing Time-of-Flight Secondary Ion Mass Spectrometry (ToF-SIMS) and the other X-ray Photoelectron Spectroscopy (XPS), are standard tools used by materials scientists—usually those developing semiconductors for electrical grids and parts for electronic devices. Like FTIR spectrometers, ToF-SIMS and XPS spectrometers bombard a material sample with energy in order to generate feedback that can be analyzed and used to identify the material. Unlike FTIR, however, ToF-SIMS and XPS work by sending an energized stream of atoms or X-rays at a sample, respectively—the feedback of which contains the detailed chemical profile of the tested material. This feedback appears on a computer screen as an undulating wavelength that scientists can learn to decipher.

"ToF-SIMS is sensitive, and more qualitative in nature, telling us whether a chemical is present or not—but not how much of it there is," Snyder explained. "XPS still doesn't give us an exact amount of what's present, though it does pick up on the more fine-tuned percentages of each chemical present on the surface of the tested material. We know that if we can find a detailed chemical profile on a material's surface using XPS, the chemicals are probably present in an amount comparable to what we'd find in the environment—which is to say, small but notable amounts."

To begin, Snyder ran each of the one hundred microbeads Mason had delivered to the University at Buffalo's chemistry lab through an FTIR spectrometer. This testing confirmed the microbeads were indeed made of plastic, mostly polyethylene and polypropylene, the first step in the chemical analysis. Hunched over the lab bench, Snyder used steel pincers to individually pick up and closely inspect ten of the plastic orbs, each the size of a grain of sand. She turned them over and over under a light and magnifier to locate any areas of flatness, however minuscule, as the larger spectrometers would only yield a signal from an even surface. With some double-sided tape and a steady hand, she positioned the beads, one at a time, on the scanning platform of the ToF-SIMS spectrometer, then the XPS machine. She then repeated the process with several microbeads she had extracted from personal care products.

Each scan took ten to twenty minutes, but the prep of each microbead took considerably longer. After scanning the microbeads, she turned to the data spun out by each machine: colorful lines on a computer screen that sloped like the undulating mountainous terrain of the nearby Allegheny Plateau. Deciphering the meanings of these peaks and valleys would take her the better part of a year.

Snyder found the surfaces of the tiny microbeads were impregnated with at least two chemicals of concern: silicone, as well as perfluoroalkyl and polyfluoroalkyl substances, also called PFAS. Silicones, long used as additives in boat paints to prevent fouling and for industrial purposes like greases, foams, and rubbers, are widespread in freshwater and marine environments where ship traffic and industrial outflows are substantial. PFAS are also found in many materials and items made and used by people, and as a result are commonly detected in aquatic ecosystems—especially those located in and around industrial corridors with factories producing products like those that have long dominated Western New York. PFAS serve as a key ingredient in firefighting foams (including those used at airports and by the military in regular training drills), and are used to coat nonstick, stain-repellant, and waterproof products—including cookware, clothing, and food packaging. PFAS are also found in grease-resistant food packaging and paints.

The microbeads pulled from the cleansing products revealed traces of silicones, but not PFAS, which made sense: Besides being used industrially, silicones are commonly added to cleansers and soaps to help increase their ability to cut through grease. "Finding them in these products shows that the microbeads they also contain are likely to have absorbed silicones before they get into waterways," said Snyder. "Microbeads are probably acting like silicone-delivery devices in this way."

While silicone exposure is linked to immune system problems, as well as certain cancers, PFAS are considered much more dangerous—and are thus a chief concern among public health experts today. PFAS chemicals, a class of more than five thousand man-made compounds containing some constellation of carbon and fluorine, have been manufactured about as long as plastic has. The bonds holding PFAS molecules together, like those that bind plastic molecules, are tight. In fact, they are bound so tightly that they do not degrade once created and instead accumulate increasingly in our bodies and the environment over time. That's why PFAS are nicknamed "forever chemicals."

A little fluorine, in the form of fluoride, is fine—even necessary—for our bodies to function. Exposure to elemental fluorine, however, can sicken us, harming bones, teeth, kidneys, nerves, and muscles; causing cancer; as well as impairing the way our hormones regulate our metabolisms and immune systems, causing obesity, early onset puberty, and reproductive issues. That PFAS persist in the environment and our bodies, rather than break down, puts us at risk of such poisoning. Analyses of blood samples drawn from humans and nonhuman animals (including polar bears, seals, and beluga whales living in the remote Arctic) suggest it's likely that most, if not all, living beings on Earth carry PFAS and a slew of other man-made chemicals in their bodies.[21] We are exposed in the products we buy, air we breathe, soil we sow, food we eat, water we drink, and wombs from which we are born.

This is the legacy of industry, and the products it makes and sells to us. Once toxic chemicals escape into the environment, they're extremely difficult to recapture or clean up. Regulators can cordon us off from highly contaminated industrial areas by delineating Superfund sites, which are meant to one day be remediated.[22] But these chemicals do not remain within the arbitrary dividing lines we draw between clean and contaminated, safe and unsafe. They travel with wind and water, covering great distances in the atmosphere, oceans, downstream, and underground. Lakes, rivers, and other freshwater ecosystems are especially prone to pollution of all kinds because of their typical proximity to both people and industrial plants.[23]

PFAS are detectable in communities' drinking water all over the world, and PFAS are known to harm human health, yet regulated and recommended exposure limits exist only in select countries. The nations that do regulate PFAS do so unevenly. While in 2020 the EU published a plan to ban most uses of PFAS across its member nations, US agencies struggled to agree on acceptable human levels of PFAS exposure. In the US, the EPA recommends—but does not require—that municipal tap water suppliers keep PFAS concentrations below seventy parts per trillion, a minute quantity significantly lower than the previous recommendation of four hundred parts per trillion. The US Centers for Disease Control and Prevention suggests the EPA's recommendation be trimmed even further, to somewhere between seven and eleven parts per trillion.[24] Meanwhile, PFAS largely continue to be manufactured around the world and, even where regulated, already exist in many of the products we use daily and have seeped into the very planet we live on. Further, regulation of PFAS and other chemicals rarely takes into account the increased likelihood of harming people more vulnerable to the detrimental effects of toxins, like children, women, the elderly, those living in underserved communities, and people with preexisting health conditions. Most chemical regulations have been historically formulated based on risks faced by "the reference man": a healthy, white, young adult male, Mason mentioned.[25]

Recommended exposure limits in tap water in Europe are similar to the previous EPA recommendation, at 500 parts per trillion, while Japan has set no suggested limits.[26] Other dangerous things you might find in your water, like bacteria and metals, are more routinely studied, tested for, and removed accordingly. That's largely because they're easier to flush from the water supply than many pervasive, long-lasting industrial chemicals like PFAS, which often require expensive filtration technology to remove.[27]

Imagine you're holding seventy grains of sand. Now throw the sand into an Olympic-sized swimming pool. Those seventy specks of sand might not be too noticeable in a large pool of water. But if you were to replace the sand with seventy toxic PFAS molecules, and then drink from the water in the pool, you could be in trouble—despite this level of contamination falling within EPA recommendations.[28] But you'd be in even more trouble, Gardella said, if the water you're drinking also contains microplastic, "which might be leading us to underestimate the effect of PFAS in water systems—you'd be exposed to what could be a significant additional amount of PFAS in that microplastic."

Besides containing industrial toxins, levels of which tend to vary based on a water supply's proximity to industry, most municipal drinking water systems globally *do* contain microplastic, and, it seems, also nanoplastic. Many of these particles are simply too small to catch with standard filters before being piped to people's taps, and what microplastic is found isn't tested for toxic chemicals, Mason explained. Loaded, the plastic pieces hide in plain sight.

In 2014, a crisis in Flint, Michigan, brought the issue of drinking water contamination to the fore of the American consciousness. There, feeble management of the city's water supply caused aging lead pipes to leach and poison water consumed by more than one hundred thousand people.[29] Yet while replacing lead pipes—which Flint's government eventually began to do, albeit at a snail's pace—can prevent more tragic lead poisoning, it cannot protect people from long-term exposure to the sneaky man-made chemicals that have for decades infiltrated the ground and surface waters we drink. PFAS, silicones, benzene, styrene, vinyl chloride, glyphosate, dioxins . . . the list of drinking water contaminants goes on and on and includes industrial chemicals, pesticides, dyes, pharmaceuticals, and the byproducts of burning oil, gas, and coal.

Though these chemicals can sicken and kill us over time, it's expensive and time consuming—and sometimes impossible—to filter them out of drinking water. Some water companies fail to catch all contaminants, because certain chemicals evade or resist treatment. Other companies simply don't try to catch them all because they're not required by federal or state law to do so—and because it's cheaper for them, that way. If you draw your drinking water from a well, to stay safe it's on you to rid your water of chemicals with an array of complex (often costly) treatment methods and regularly changed filters.[30]

When our human ancestors were hunting and gathering eleven thousand years ago, a giant jagged ice sheet that had frozen over most of what is now Canada and the northernmost US began melting and retreating as Earth's most recent glacial period came to a close. As it melted, the shrinking ice sheet incrementally exposed five enormous gashes in the Earth, which it had inflicted upon its creation tens of thousands of years earlier. Over thousands more years, that melted ice—supplemented by precipitation, rivers, and underground springs—filled the terrestrial scars that would hold what we know today as the Great Lakes.[31]

Today, more than 10 percent of the US population and more than 30 percent of the Canadian population—a total of forty million people—regularly drink water pulled from the Great Lakes.[32] Communities around all the Great Lakes—but particularly those home to a greater number of industrial plants, Lakes Michigan, Huron, and Erie—have been exposed to elevated levels of PFAS in their municipal drinking water. These same people are also exposed to microplastic, mostly synthetic fibers, according to research done by Mason and others.[33]

"We need to keep looking at the surfaces of microplastic particles found in the Great Lakes to understand what they contain and how that might affect the whole food chain," Gardella told me. "We need to understand how these microplastic particles might have exposed fish, plants, and people to the chemicals they carry along with them." What needs to come next, Snyder added, is determining to what degree toxins stick to the little plastic pieces—how much poison is contained in each particle. She's working on figuring that out now.

Besides providing people with drinking water, Lake Erie holds more than half of all fish collectively found in all the Great Lakes. With less water to heat, Lake Erie warms faster than its counterparts, fostering the abundant algae growth on which zooplankton, like daphnia, feed.[34] The abundant daphnia provide sustenance to larger-sized zooplankton like the shrimp-like gammarus, as well as baby fish belonging to dozens of species commonly caught and eaten by people. Daphnia are one of the earliest foods that baby whitefish, perch, shad, bass, and rainbow trout sink their teeth into.[35] We catch, kill, and then sink our teeth into these fish once they've reached maturity.

Since there's no longer any question whether or not daphnia are consuming microplastic and whether or not it's harmful, understanding the ecosystem-level effects of their plastic predilection, in nature, is a critical next step scientists need to address. And this must be done within the context of the many other problems that humans cause for the nonhuman life with which we share the planet. So, starting at the foundation of the Great Lakes' ecosystem is a logical place to begin.[36]

"Ultimately, if daphnia are ingesting microplastic, they can deliver it—and the chemicals it carries—into other parts of the food web," said Wigdahl-Perry. "The only way to know what's happening is to put all the pieces together while trying to also understand each issue individually. This is why collaboration is so important—we can all bring in-depth answers to the table."

Daphnia already face numerous threats to their existence. Factories and factory farms expel pollution that feeds algae blooms that rob Lake Erie's waters of oxygen and increase levels of harmful pollutants. Climate change has caused Lake Erie's water level to sink and its temperature to rise. And to make way for development, over the past two hundred years, people have destroyed 95 percent of Lake Erie's shallow wetlands—areas that provide critical habitat for wildlife and also protect human communities from floods and storms, in addition to improving the lake's overall water quality.[37] On top of all this, could microplastic be the straw that breaks the daphnia's back?

"If all daphnia were to die off in Lake Erie, I can say with certainty you'd see changes, potentially all the way to the top levels of the food web," Wigdahl-Perry said. "The good thing about natural ecosystems is that when one species dies off, another will jump in to take its place.

Depending on what takes the daphnia's place, there may not be a big swing in the food web as a whole. But any time you upset the balance of nature, you open the door for non-native species, which are opportunistic by nature and can expand to take over any open spaces faster than native species can."

Already, Lake Erie is home to more than 185 non-native species, introduced through ship travel and ballast and on the plastic constantly moving through the Great Lakes system. The spiny waterflea, a carnivorous zooplankton native to Europe and Asia, is one invader that's managed to colonize not just Lake Erie but all of the Great Lakes, where it feasts on daphnia and other native zooplankton in significant quantities.[38]

There is currently no way to oust the spiny waterflea from the Great Lakes. They are too small and too numerous. Pulling these opportunistic creatures out of the lakes without harming native plants and animals would be near impossible, just as it would be similarly hopeless to try to rid the lakes of microplastic and nanoplastic particles. Spiny waterfleas' presence and predilection for eating native zooplankton has already started to shift the balance of Lake Erie's web of life: The lake's populations of native crustaceans and perch—which have no taste for the invasive zooplankton—are presently plummeting.[39] The indigenous zooplankton they rely on eating for survival are being picked off by plastic and the invaders the material carries, Wigdahl-Perry explained.

Maybe over time daphnia and other animals could mostly adapt to living in a world teeming with microplastic. But what happens when you add other stressors such as climate change, dumping of toxic chemicals, and habitat loss? The picture looks quite grim. "Looking at these multiple factors will either fascinate you or make you want to pull your hair out," as Wigdahl-Perry put it.

With so many threats to wildlife, we humans shouldn't expect to be exempt from the deleterious effects of plastic.

# CHAPTER 7
## The Plastic Within Us

Back across the Atlantic Ocean, in Denmark, I'd heard that scientists from Aalborg University had begun to explore the impacts of microplastic on us humans. Aalborg is the fourth-largest city in Denmark, located on a large peninsula called Jutland. A five-hour train ride from Copenhagen brought me directly to the university, which lies south of a retired limestone quarry now filled with alkaline water as turquoise as that in the tropics, and adjacent to a patchwork of green-and-brown farmland.

Alvise Vianello, a postdoc at the university, sprang around the big-windowed halls of its Department of Civil Engineering on the toes of his running shoes, pointing out to me various projects in progress: "This is a filter being tested to catch microplastic in sewage, because in Denmark and other countries, sewage sludge mixed with microplastic is routinely spread on agricultural fields. . . . This is a new FTIR machine for identifying various types of microplastic, even the very smallest pieces. . . . This is sediment collected from riverbeds in Norway, which seem to have a lot of microplastic in them." The spry Italian scientist ducked between whizzing and whirring machines, disassembled mechanisms and coils of tubing, petri dishes dotted with plastic fibers and sample jars brimming with water and mud. Swiftly rattling off his department's many research endeavors, Vianello came up for air only for the briefest of questions and to exuberantly wave hello to any lab coat–clad peers who passed. He finally slowed a bit upon reaching the university's cavernous basement. Its main feature was a pool-sized tank, one end of which was banked by smooth, dark pebbles. The tank, Vianello explained, was used to test how much energy can be generated by—and possibly harnessed from—waves in the North Sea and Danish fjords.

Beyond the pool, a beige-colored mannequin sat at a table behind an array of hoses, boxes, and freestanding shelves. A few light raps with my knuckles revealed sounds indicative of substance: the clang of metal and thud of resin. The nubby end of a copper tube drooped from what appeared to be each eye socket, above a mouth punched open in a small "o." The mannequin's knees, made of flexible tubing, bent beneath the table, and its thick arms, lacking hands, hung straight by its sides. From the back of its head a copper tube curved down toward the floor. Where two feet may have existed was one wide one, planted flat on the floor below. This lent the crudely human mannequin the fixed appearance of someone who might have been staring down at something surprising.

"So, this is my project," he announced with delight, gesturing toward the mannequin, which resembled a simpler, less sparkly version of *Star Wars'* C-3PO. But in some ways, this mannequin was even more humanlike than the famous robot: It could breathe.

During his experiments, Vianello had heated the mannequin until it nearly reached a human's body temperature. Then, he flicked on the mannequin's artificial lungs, two cylinders driven by pistons connected to an electric motor. This triggered the mannequin to draw in air, which moved through a series of interior tubes equipped with fine silver filters to catch any particles that might be in the air. After circulating inside the mannequin, each breath in was released from the tubes out through the mannequin's nose—the two copper tubes I'd mistaken for its eyes.[1]

The mannequin took about ten to twelve breaths per minute, the same rate a healthy, resting adult might, Vianello said. At the end of

2017, Vianello had brought the mannequin to life in three Danish college students' apartments, three times in each apartment for twenty-four hours at a stretch. After the inanimate houseguest enjoyed its stay in each of the apartments, Vianello analyzed the microplastic the mannequin captured in its lungs with the help of Aalborg's lead plastic expert, Jes Vollertsen; Aalborg professor Rasmus Lund Jensen; and Tsinghua University researcher Li Liu.[2]

After meeting the mannequin, Vianello and I walked up to his office to meet Vollertsen and take a look at the team's results, which, at the time, were soon to be published in the journal *Scientific Reports*. On a projector screen, Vianello and Vollertsen revealed a color-coded map of all the particles the mannequin had inhaled during one of its twenty-four-hour apartment-sitting sessions. Light gray blobs indicating the presence of protein dominated the map—in this case, skin cells—while there were fewer dark gray blobs and threads indicating plant material. The gray shapes were interspersed with a rainbow of blobs and lines representing dozens of types of plastic fragments and fibers.

"From what we can tell, it's possible people are breathing in around eleven pieces of microplastic per hour when indoors," Vianello explained. An FTIR analysis revealed that the most common type of plastic the mannequin breathed in was polyester, which is probably constantly shedding off the furniture, carpets, and clothing we keep in our homes.[3]

We are no better protected from plasticized air outdoors than we are indoors. Plastic fibers, fragments, foam, and films are perpetually floating into and free-falling down on us from the atmosphere. Rain flushes micro- and nanoplastics out of the sky back to Earth. Plastic-filled snow is accumulating in urban areas like Bremen, Germany, and remote regions like the Arctic and Swiss Alps alike.[4]

Wind and storms carry particles shed from plastic items and debris through the air for dozens, even hundreds, of miles before depositing them back on Earth. Dongguan, China; Paris, France; London, England;

and other metropolises teeming with people are enveloped in air perpetually permeated by tiny plastic particles small enough to lodge themselves in human lungs.[5]

Urban regions are especially replete with what scientists believe could be one of the most hazardous varieties of particulate pollution: plastic fragments, metals, and other materials that have shed off synthetic tires as a result of the normal friction caused by brake pads and asphalt roads, and from enduring weather and time. Like the plastic used to manufacture consumer items and packaging, synthetic tires may contain any number of a manufacturer's proprietary blend of poisons meant to improve a plastic product's appearance and performance.[6]

Tire particles from the world's billions of cars, trucks, bikes, tractors, and other vehicles escape into air, soil, and water bodies. Scientists are just beginning to understand the grave danger: In 2020, Washington State researchers determined that the presence of 6PPD-quinone, a byproduct of rubber-stabilizing chemical 6PPD, is playing a major factor in a mysterious long-term die-off of coho salmon in the US Pacific Northwest. When Washington's fall rains herald spawning salmon's return from sea to stream, the precipitation also washes car tire fragments and other plastic particles into these freshwater ecosystems. In recent years, up to 90 percent of all salmon returning to spawn in this region have died—a number much greater than is considered natural, according to local researchers from the University of Washington, Tacoma. As University of Washington environmental chemist Zhenyu Tian explained in an interview with Oregon Public Broadcasting, 6PPD-quinone appears to be a key culprit: "You put this chemical, this transformation product, into a fish tank, and coho die really fast."[7]

While other researchers have previously searched for, and detected, microplastic dispersed in indoor and outdoor air, Vianello's study was the first to do so using a mannequin emulating human breathing via mechanical lungs.[8] Despite the evidence his research provides—that plastic is getting inside of human bodies and could be harming us—modern health researchers have yet to systematically search for it in people and comprehensively study how having plastic particles around us and in us at all times might be affecting human health.

Vianello and Vollertsen explained that they've brought their findings to researchers at their university's hospital for future collaborative research, perhaps searching for plastic inside human cadavers. "We now have enough evidence that we should start looking for microplastic inside human airways," Vollertsen said. "Until then, it's unclear whether or not we should be worried that we are breathing in plastic."

He speculated that some of the microplastic we breathe in could be expelled when we exhale. Yet even if that's true, our lungs may hold onto much of the plastic that enters, resulting in damage.

Other researchers, like Joana Correia Prata, a PhD student at the University of Aveiro in Portugal, have highlighted the need for systematic research on the human health effects of breathing in microplastic. "Microplastic particles and fibers, depending on their density, size, and shape, can reach the deep lung causing chronic inflammation," she said. People working in environments with high levels of airborne microplastics, such as those employed in the textile industry, often suffer respiratory problems, Prata has noted. The perpetual presence of a comparatively lower amount of microplastics in our homes has not yet been linked to specific ailments.[9]

While they've dissected the bodies of countless nonhuman animals for decades, it's only been a few years since scientists began exploring human tissues for signs of nano- and microplastic. This, despite strong evidence suggesting plastic particles—and the toxins that adhere to them—permeate our environment and are widespread in our diets. In the past decade, scientists have detected microplastic in the bodies of fish and shellfish; in packaged meats, processed foods, beer, sea salt, soft drinks, tap water, and bottled water. There are tiny plastic particles embedded in conventionally grown fruits and vegetables sold in supermarkets and food stalls.[10]

As the world rapidly ramped up its production of plastic in the 1950s and '60s, two other booms occurred simultaneously: that of the world's human population and the continued development of industrial agriculture. The latter would feed the former and was made possible thanks to the development of petrochemical-based plastics, fertilizers, and pesticides. By the late 1950s, farmers struggling to keep up with feeding the world's growing population welcomed new research papers and bulletins published by agricultural scientists extolling the benefits of using plastic, specifically dark-colored, low-density polyethylene sheets, to boost yields of growing crops. Scientists laid out step-by-step instructions on how the plastic sheets should be rolled out over crops to retain water, reducing the need for irrigation, and to control weeds and insects, which couldn't as easily penetrate plastic-wrapped soil.

This "plasticulture" has become a standard farming practice, transforming the soils humans have long sown from something familiar to something unknown. Crops grown with plastic seem to offer higher yields in the short term, while in the long term, use of plastic in agriculture could create toxic soils that repel water instead of absorbing it, a potentially catastrophic problem.[11] This causes soil erosion and dust—the dissolution of ancient symbiotic relationships between soil microbes, insects, and fungi that help keep plants alive.[12]

From the polluted soils we've created, plants pull in tiny nanoplastic particles through their roots along with the water they need to survive, with serious consequences: An accumulation of nanoplastic particles in a plant's roots diminishes its ability to absorb water, impairing growth and development. Scientists have also found early evidence that nanoplastic may alter a plant's genetic makeup in a manner increasing its susceptibility to disease.[13]

Based on the levels of micro- and nanoplastics detected in human diets, it's estimated that most people unwittingly ingest anywhere from thirty-nine thousand to fifty-two thousand bits of microplastic in their diets each year. That number increases by ninety thousand microplastic particles for people who regularly consume bottled water, and by four thousand particles for those who drink water from municipal taps.[14]

In 2018, scientists in Austria detected microplastic in human stool samples collected from eight volunteers from eight different countries across Europe and Asia.[15] Clearly, microplastic is getting into us, with at least some of it escaping through our digestive tracts. We seem to be drinking, eating, and breathing it in.

A few scientists, including Kristian Syberg, have recently uncovered another potential consequence of plastic exposure, one particularly relevant to our modern human society freshly struck by a devastating pandemic: Harmful viruses and bacteria have a tendency to colonize plastic particles and objects, which are not easily cleaned like other materials such as glass and metal. The same spongelike surfaces that make plastic attractive to toxic chemicals also attract microbes. This could mean plastic and its particles may be capable of spreading disease. In Zanzibar, an archipelago off the coast of Tanzania, in East Africa, Kristian and several of his colleagues from Roskilde University detected cholera, salmonella, and *E. coli* on plastic debris found littered in communities where these illnesses are known to circulate. While doing their research, Kristian and his team noticed Zanzibar's street vendors sold hand-pressed sugarcane juice from plastic bottles. When asked, the juice sellers told the researchers they'd simply collected, rinsed, and refilled plastic bottles pulled from the piles of waste all around, the same contaminated trash the team had tested.[16]

We don't know yet exactly what those plastic particles do while inside us, if they are a significant contributor to the spread of diseases that might be hiding invisibly on their surfaces, or which chemicals may linger in our bodies long after plastic has passed through. But we can make an educated guess: In a world completely permeated with plastic and toxic chemicals of our creation, we have a fate akin to that of all Earth's other creatures.

Each piece of plastic possesses a proprietary chemical composition;

each carries with it, and carries on, plastic's toxic legacy. Scientists have demonstrated that when wild and laboratory animals like fish ingest microplastic, they also get a dose of the toxic chemicals microplastic carries.[17] And these chemicals are linked to cancers, reproductive problems, metabolic disorders, autoimmune diseases, malnutrition, and other health issues—in both people and other animals. Yet we humans rarely stop to consider our vulnerability to plastic, a substance that is sickening and killing albatrosses and whales, dolphins, fish, and countless other creatures right before our eyes. Perhaps we have hesitated to search inside ourselves because we are afraid of what we might find.

Scientists are just beginning this search in earnest. In August 2020, a group of Arizona State University researchers, led by Rolf Halden, director of the university's Center for Environmental Health, announced at a virtual meeting of the American Chemical Society that his research team had discovered both plasticizer chemicals and basic plastic compounds, called monomers, in dozens of samples of donated human lungs, livers, spleens, and kidneys. BPA, a chemical known to harm the developing brains and bodies of children and widely added to plastic since the 1960s, was found in all of the human tissues sampled. But they stopped short of identifying actual pieces of nano- and microplastic in the tissues. In separate experiments, Halden and his team spiked human tissue samples with plastic particles to test if a tool, called a flow cytometer—which scans individual cells using a light beam, revealing physical and chemical properties—could help locate them. Other researchers have applied flow cytometry to plastic pollution research, specifically to detect plastic particles suspended in freshwater and seawater samples. According to Halden, the logical next step is to apply flow cytometry to find microplastic in the landscape of our bodies.[18]

"It would be naive to believe there is plastic everywhere but just not in us," Halden told the *Guardian* in August 2020. "We are now providing a research platform that will allow us and others to look for what is invisible—these particles too small for the naked eye to see. The risk [to health] really resides in the small particles."[19]

Just months after Halden and his team released their findings, a group of Italian scientists revealed they had documented plastic particles in the womb, for the first time in human history. The researchers, affiliated with two universities and two hospitals, created a "plastic-free protocol," to which medical teams adhered for the delivery of six healthy women's pregnancies. The scientists took small tissue samples across the maternal and fetal sides of each placenta, and from the amniotic sacs that held the babies' developing fetuses. In total, four of the six placentas contained microplastic across all areas inspected. The tissue samples, representing just 4 percent of each placenta, collectively contained twelve plastic particles dyed blue, red, orange, and pink—suggesting they were shed from common plastic items like cosmetics and personal care products. While the women had normal births, the scientists expressed their concern that it seems at least some plastic particles circulate the human bloodstream after being ingested or inhaled.[20]

"With the presence of plastic in the body, the immune system is disturbed and recognizes as 'self' (itself) even what is not organic," said Dr. Antonio Ragusa, who was involved in the study and works as director of obstetrics and gynecology at Fatebenefratelli Hospital in Rome, in an interview with Italian newspaper *la Repubblica* in December 2020. "It is like having a cyborg baby: no longer composed only of human cells, but a mixture of biological and inorganic entities. The mothers were shocked."[21]

If tiny plastic particles can circulate human bloodstreams, as the placental research suggests, micro- and nanoplastics may also accumulate in our bodies while leaching poisons that harm our immune systems and put us at increased risk of serious disease.[22] Informing this hypothesis is a large and widely understood body of research probing the health effects of one of the deadliest known forms of particulate pollution on the planet: soot, or black carbon—the fine airborne matter that's emitted during incomplete combustion of hydrocarbon-based substances, like wood and coal. Like plastic particles, soot is laced with all manner of toxic chemicals, some inherent and others acquired while in nature. And just as microorganisms like viruses and bacteria adhere to plastic, they are also known to adhere to soot particles. When inhaled, soot lodges deeply in the lungs and is absorbed by the bloodstream. Soot particles have been detected in and are known to adversely—sometimes lethally—affect human brains, skin, hearts, lungs, and other organs.[23]

Plastic particles, like soot, are not monolithic and so tend to resist scientific classification into neat categories. One of the biggest challenges plastic pollution scientists face is finding a way to standardize and systematize the identification of plastic particles in our bodies and the environment, a fundamental element of research on other toxic pollutants. FTIR spectroscopy, the widely used microplastic research technique Torsten employed to analyze our samples from the Pacific and Sam Mason used to scan samples pulled from Great Lakes sediments, can tell scientists what type of plastic a particle is made of. That's helpful, but as I learned from Abigail Snyder and Joseph Gardella Jr. at the University at Buffalo, FTIR is incapable of elucidating particles' often complex chemistries with precision. Each piece of plastic may contain any potential constellation of toxins linked to various health risks.

In 2019, a team of Woods Hole Oceanographic Institution researchers presented a possible way forward when they pointed out that, just a few decades ago, scientists resolved a similar research issue when studying soot particles, which also vary in chemistry and appearance. Air pollution researchers established a continuum of characteristics, from particle size and color to a range of chemical parameters, to help discern what is soot and what isn't, and what risks each particle might carry—physically and chemically.

"On their face, all plastics share similar characteristics: They are carbon-based materials with varying degrees of hydrogen, oxygen, and nitrogen (from additives and colors)," the Woods Hole scientists explained. "However, the usefulness of the overly simplistic, all-encompassing term 'plastics' ends quickly when we look more broadly at the chemical composition of these specially engineered materials and the forces that act on them throughout their life cycle. In reality, plastics come in all shapes, sizes, and proprietary formulations designed for largely different yet dedicated purposes. They are not the same as pollutants like . . . [DDT] or lead, which have robust methods and a common language to communicate among fields about how to measure and report concentrations of these pollutants and the threats they pose to life. Consequently, the growing field of environmental plastics is facing a demanding task of learning to differentiate among the infinite combinations of size, shape, and formulations, resulting in an indefinite timeline for us merely to be able to understand what it is we mean when we talk about plastics."[24]

Today plastic pollution research is moving forward fast. Fortunately, researchers in this expanding scientific field have not had to start completely from scratch: Scientists are now adapting existing research tools and methods, including those long used by industries to analyze material samples and study the effects of soot—another particulate pollutant. Scientists are closer than ever to fully understanding plastic's distribution in our environment and our bodies and its seemingly wide range of detrimental effects.

While the work continues to identify the plastic particles in our bodies and illuminate their impacts, some of the health risks of plastic are already established—but they happen before the plastic is first shipped or sold.

Denka Performance Elastomer in the rural, predominantly African American community of Reserve in St. John the Baptist Parish, Louisiana, March 2020. Photo by Erica Cirino.

# PART III

# People and the Plastic Industry

## CHAPTER 8
# Welcome

"We have our community, we have our lawyers, we have God. But no one we've elected is helping us oppose the petrochemical industry that's destroying our parish. We show up, and pray the courts will deliver justice." At the front of the auditorium, a woman with sparkling nails and short-shorn hair spoke quietly but confidently. Like many of the neighbors seated at long tables before her, she wore a yellow tee emblazoned with a black fist rising above an array of human silhouettes.

This was Sharon Lavigne, then sixty-eight years old, a retired special education teacher and lifelong resident of Welcome, Louisiana. Welcome is a tight-knit community of around eight hundred people located in St. James Parish on the west bank of the Mississippi River, about sixty miles west of New Orleans. It was early March 2020, and Lavigne had invited her neighbors to gather for a midday meeting at a Catholic church in Convent, the parish seat. From the auditorium's kitchen, the heady odor of simmering gumbo wafted out from the kitchen, where volunteers prepared a lunch for the community. On the agenda was a pressing community matter: the planned construction of an enormous plastic factory in their neighborhood, which was already surrounded by busy railroad tracks frequented by trains carrying toxic chemicals, an ammonia plant, a polystyrene factory, oil and gas pipelines, and more than fifty enormous chemical storage tanks, some situated less than one thousand feet from the nearest homes.

Lavigne's goal that day was straightforward: "To remind St. James to keep faith, and keep showing up, because our lives depend on it."

Living in Welcome, a predominantly African American neighborhood, Lavigne has watched her neighbors and family members grow ill with cancers, heart problems, autoimmune disorders, and other conditions known to be caused by a variety of factors that include, most notably, exposure to industrial pollutants. The governmental, political, and corporate systems and cultures enabling and favoring the unjust placement of industry in Black communities like Welcome are a textbook example of environmental racism.

The consequences are clear: In America, Black people are more likely to die prematurely due to toxic air pollution, because decades of racist policies have made sure the air they breathe is the dirtiest in the country.[1] People who are Indigenous, unhoused, low income, or belong to other historically underserved groups share this increased risk of pollution exposure. While America's Clean Air Act of 1970 has over time reduced racial disparity in populations made to bear industry's pollution burden, to this day industries continue to overwhelmingly target communities of color.[2]

Protecting communities requires constant effort. In 2015, Lavigne first spoke out against development when she learned the local government had covertly changed land-use rules, effectively fast-tracking construction permits for two large petrochemical plants in St. James. In 2018, she and her neighbors watched as the state wrested an enormous rural plot, formerly home to two plantations, from the local community and sold it to FG LA LLC (FG), one of many business endeavors run by the enormous Taiwanese manufacturing conglomerate Formosa

Plastics Group. When that happened, Lavigne—who has seen some local businesses and residents depart St. James as industry has moved in—retired from teaching to form a Christian faith–based activist organization, called RISE St. James. Lavigne and RISE's other Black community leaders, such as Barbara Washington and Stephanie Cooper, call on RISE's members and allies from other environmental organizations to help document pollution in communities, attend public hearings and community meetings, bring lawsuits against polluters and dysfunctional government regulatory agencies, write to lawmakers, and pray.

The previous year, RISE had focused on, and ultimately succeeded in, pushing away a Chinese chemical company called Wanhua. The company had planned to build a $1.25 billion chemical factory in St. James that would make methylene diphenyl diisocyanate (MDI), a key—and highly dangerous—ingredient used to produce polyurethane foam. At the time of that March 2020 meeting, RISE was focused on stopping Formosa's planned ten-year construction of a 2,400-acre, $9.4 billion plastic factory called the Sunshine Project. Before construction began, forecasts projected that the Sunshine Project complex, slated to produce an array of petrochemicals and various types of raw plastic, when up and running would release substantial amounts of plastic and carcinogenic chemicals into Welcome. If built, the Sunshine Project would become one of the largest plastic factories in the world.

According to a ProPublica investigation, emissions from the Sunshine Project would more than triple Welcome residents' exposure to cancer-causing chemicals and double the risk of such exposure for people living across the river in Convent. Formosa's plant would also spew more than 13.6 million tons of carbon dioxide into the air each year, the amount of carbon three million additional cars would add to the atmosphere if each driven for one year.[3] This, in a region already feeling the heat cast by humanity's uncontrolled greenhouse gas emissions— oppressively hot, humid days and sinking coastlines, deadly floods and catastrophic storms, all occurring more frequently and intensely than ever before. Even if the US and rest of the world curbed carbon emissions right now, between 2040 and 2060, experts estimate at least 5 percent of St. James will be regularly underwater at high tide; temperatures soaring high enough to make it too dangerous to go outside will become a normal occurrence.[4]

While plastic products pollute our environment and—as scientists have detected—our very bodies, they begin inflicting harm before they're sold, much farther up the pipe. A survey of plastic's impacts isn't complete without taking into account the ways in which plastic manufacturing pollutes air, soil, and water—especially in communities of color, like Welcome.

The people of St. James already live within a ten-mile radius of twelve petrochemical refineries, in addition to other massive industrial facilities emitting high levels of pollution—including an ammonia plant and a steel factory. Surrounded by industry, communities in St. James are already frequently subject to boil-water advisories due to unsafe levels of chemicals and bacteria in their drinking water, which is pulled locally from the Mississippi River. Deadly and injurious explosions, accidents, petroleum and petrochemical leaks, and fires plague industrial complexes situated near schools, homes, and businesses. If not completely displaced by industry, the often-Black neighborhoods next to which factories are erected, like Welcome, are separated from adjacent industries with nothing more than a fence—a characteristic of these neighborhoods that's given rise to the moniker *fenceline community*.

Few know more about the history of fenceline communities in the region than Craig E. Colten, whom I met prior to RISE's meeting, in his book-filled office at Louisiana State University (LSU). At LSU, Colten works as a historical geographer, author, and professor, focusing on his home state of Louisiana. He's written extensively about its racially charged and industrially developed landscape, particularly in the wake of Hurricane Katrina. And as a local who came of age in the '60s and '70s, Colten has had firsthand experience with Louisiana's oil and gas industry explosion—though back then, he, unlike many of his counterparts, chose not to work in the oil fields and on the rigs.

When asked to explain how industry came to dominate many of Louisiana's rural, formerly agricultural communities, like Welcome in St. James, Colten leaned back in his desk chair, folded his hands on his lap, and dove into a brief history lesson: Following the abolition of slavery in 1865, many freed Black people continued working as oppressed laborers, tenant farmers, or sharecroppers on the same white-owned plantations where they'd been subjected to slavery. But over time, many Black farmers could afford to release themselves from these abusive labor systems, staking claims of ownership on plots of sprawling green fields, often on the very same grounds where their parents and grandparents had been enslaved. Black homesteaders across Louisiana and the rest of the American South shaped new communities where they could finally work on their own farms, on their own terms.

At the same time these Black communities emerged, American industrialists—empowered by the discovery that crude oil could be burned for energy and processed into various chemicals—redoubled their efforts to revolutionize the daily lives of the wealthy with fossil fuels. Refineries were assembled all over the country, particularly in regions where crude was readily tapped and infrastructure was easily built along waterways. Louisiana, awash in oil, saw construction of its first refinery in 1909 when Standard Oil set up shop in Baton Rouge. Vastly expanded and now owned by ExxonMobil, the refinery still stands today as the fifth-largest petroleum refinery in the United States and second largest in Louisiana.[5]

"In Louisiana, natural levees near the Mississippi were chosen to become a part of the petrochemical complex," Colten explained. "These wide swaths extended miles out from the river and encompassed plantation land. Oil companies bought up the plantations and began developing refineries throughout the '20s and '30s, displacing those living and farming in free Black communities. World War II and the need for materials, especially synthetic rubber, sped up construction of more industry—in fact, the State of Louisiana *welcomed* industry by implementing policies friendly to big corporations."

This enthusiastic welcome continues today. In 2015, Louisiana's then governor Bobby Jindal offered Formosa a $12 million grant and other support when the company's chairman Bao-Lang Chen began scouting potential construction sites in rural St. James. At the time, the company planned to build farther downriver, in a whiter community near Gramercy Bridge. Colten explained that the white residents weren't happy with that, so Formosa silently moved its project to the predominantly African American community of Welcome.

What's left along this stretch of the Mississippi is a patchwork of sugarcane plantations and petrochemical complexes, the former with a legacy of slavery and soil degradation and the latter with a legacy of spills, explosions, and widespread pollution. As industry has closed in, breathing room has been hard to come by. Many people are fearful for their lives and would prefer to leave. Some people can afford it, most often when paid by corporations to abandon their contaminated properties. Some people in fenceline communities must hang on the fringes of their increasingly developed and polluted hometowns because they lack the means to leave.

Nearby, Louisiana's largest oil refinery, belonging to Marathon Petroleum Corporation, sprawls across 3,500 acres of land. From Louisiana Highway 44, just a dozen of the refinery's more than one hundred cylindrical storage tanks holding crude, petrochemicals, and waste are visible. Running from the refinery, thick pipes have been suspended to create overpasses through which oil and petrochemicals gush right above the highway and across a green levee to meet tankers docked in the Mississippi. The scenic route along the back roads of the refinery reveals a glimpse of a handful of hazy, green toxic ponds of tailings; a set of rusting rails beckoning trains full of oil; towering smokestacks too numerous to count while driving fifty-five miles per hour; and a frequently burning red-hot flare. This enormous complex—which produces, among many other fossil-fuel products, the petrochemicals used to make plastic—sits next-door to St. James, in St. John the Baptist Parish. It occupies land that was considered part of the former San Francisco sugarcane plantation.

Kept separate from the petrochemical complex by a barbed-wire fence, and situated along a road now closed to public traffic, a meager green plot breaks up the industrial sprawl. This is Zion Travelers Cemetery, which houses a collection of weathered cement tombs, carved crosses, and engraved headstones, some sinking into the soft Louisiana loam like ruins of an ancient city. Even in death, Black residents are forced to exist on the fenceline. Or in the case of those laid to rest in this plot, within it.

Communities throughout St. James, St. John the Baptist, and Louisiana's other "River Parishes"—those located along the Mississippi between New Orleans and Baton Rouge—shoulder some of the worst impacts of industry in the US. While most of the nation's residents live with a cancer risk of around six to twenty-five in a million, throughout this region, cancer risks run significantly higher, reaching two thousand in a million in part of St. John the Baptist Parish, where a neoprene factory, Denka Performance Elastomer, emits a constant cocktail of chemicals, including carcinogenic chloroprene gas.[6]

As a result, this region of Louisiana has acquired a grim reputation as Cancer Alley. In total, it is home to approximately 150 industrial plants—many of which produce chemicals used to make plastic—stretching across eighty-five miles of rural land, along both banks of the Mississippi. From the worst-polluted part of the Pacific Ocean, I had traced the destructive path of plastic back to a major source, in the most notoriously toxic region of America's petrochemical landscape.

On my way to Welcome from New Orleans, in Norco, St. Charles Parish, I drove around two enormous refineries, one owned by Valero and the other by Royal Dutch Shell, and past two chemical plants, on narrow roads lined with sludgy drainage ditches slick with oil. These complexes surround Norco's few thousand human inhabitants and their homes, shops, restaurants, post office, and places of worship. Norco was named by and for New Orleans Refining Company (NORCO), the town's earliest industrial inhabitant, in 1916, following its purchase of former plantation land. In 1929, Shell acquired NORCO's refinery, expanding operations significantly to include production of chemicals used to make plastic, on agricultural land I'd later learn had been wrested from the descendants of formerly enslaved African Americans who had established farms in a community called Diamond.[7]

Diamond, which began as a small Black neighborhood, has been wracked by two lethal explosions at the Shell plant, in 1973 and 1988. Its residents have long suspected that their constant exposure to toxins was making them sick, though health officials have suggested the increased incidence of cancers and other diseases their community has seen could also be caused by smoking and other lifestyle choices. Norco's white neighborhoods, located farther afield from the town's most dangerous industrial operations, are less exposed.

Diamond once was a vibrant African American community. Today four mostly empty streets remain, running through tidy plots, many barren, a few with still-occupied homes. Diamond is a modern ghost town born out of necessity, as revealed by investigations and justice-seeking efforts spearheaded by Margie Eugene-Richard, an African American woman who grew up just twenty-five feet from Shell's Diamond petrochemical plant. Having witnessed Shell's numerous disasters striking in her own backyard, and the company's pollution sickening close friends and family members, Richard spearheaded efforts to hold the company accountable. With the help of Louisiana Bucket Brigade, the Sierra Club, and other nonprofit allies, Richard formed a community group called the Concerned Citizens of Norco, which called on residents to gather air samples with "buckets": low-cost, DIY research tools typically constructed from rigid five-gallon plastic containers, tubes, valves, and Tedlar bags (which are designed to hold volatile gases). Once collected, air is sent to laboratories for chemical analyses.

In 1994, personal injury attorney Ed Masry, who worked with Erin Brockovich, equipped residents of Contra Costa County, California, with the earliest iteration of these buckets to collect polluted air in neighborhoods near Unocal Corporation's Rodeo refinery. This air-sampling effort helped reveal unchecked air pollution that had sickened thousands of people living nearby. Unocal ultimately settled an $80 million lawsuit paid out to some six thousand residents. Since, air-sampling kits used by so-called bucket brigades have helped many communities across the US keep tabs on their local air pollution levels and hold industries accountable for violating emissions regulations.[8]

Diamond residents used their air pollution data, which revealed concerning levels of toxic chemicals, to take Shell to court, demanding relocation. During many frustrating years of litigation, Shell continued to pollute. Finally, in 2000, after Richard traveled straight to Shell's top corporate officials working at The Hague, the company made its first buyout offer. But it was offensively low: just $26,000 per property. Richard and her allies kept pushing back to get a fair price for giving up their homes. Finally, in 2002, Shell offered to buy out Diamond's residents—extending home-improvement loans to the few who chose to stay—and reduce its emissions, formally acknowledging that living in Diamond was too risky.[9] Most people, including Richard, have left

Diamond, though Richard would devote her life to advocating for other communities overtaken by industry in the US and abroad.

In LaPlace, St. John the Baptist Parish, a handful of cows languidly roamed a scruffy patch of dead roadside grass at the foot of the steaming, gleaming scaffolding of yet another chemical complex. The blue-and-white logo affixed to a large chemical storage tank read "Denka Performance Elastomer." When I stepped out my car to get a better look at the animals, the sharp scent of industrial emissions stung my sinuses, and my temples began to throb. Almost immediately, I noticed a pickup truck outfitted with security mirrors and flashing lights rolling toward me. I hurriedly snapped a few photos of the cows—and inevitably, the plant—before returning to my car and driving on.

Near Garyville, approaching St. James Parish, the landscape and everything that occupied it appeared increasingly sepia toned. The streets, the fences, the houses, the electricity wires, and the grass that miraculously continued to grow—everything was acquiring a rusty tint that intensified in hue when an industrial complex came into view a few miles down the road. This one was a hodgepodge of round-topped domes, silos and pipes, and smokestacks, all coated with a layer of bauxite ore, a red claylike substance used in aluminum refining, imported from Jamaica. Bauxite dust, which often contains traces of heavy metals, is considered an occupational hazard for people who work with the ore. For miles, the clay clung to everything, even the air, which felt gritty inside my mouth. The wind carried the plant's toxic emissions, sending mercury invisibly into the air and sweeping it across the orange landscape, where it accumulated in the soil, nearby streams and rivers, and the mighty Mississippi.[10]

I continued driving past more toxic tailings ponds, more chemical plants, more piles of industrial waste, until I reached the Sunshine Bridge. After crossing the cantilever bridge, I followed River Road past the Mosaic company's fertilizer and ammonia factory and AmSty's poly-styrene plant to finally arrive in Welcome. When I arrived, I climbed up the grassy levee to take a look at the river. I could see a grain barge loading up against a collection of floating storage containers strapped together like a giant metal raft near the undeveloped bank—just grass and mud and twisting live oaks—most of them dead and crumbling. Upriver, I could see a tangle of thick pipes reaching across the levee and over the highway, supplying petroleum to yet another chemical plant. The site of the proposed plastic factory, an enormous acreage of over-grown grass, was cordoned off by a tall, chain-link fence, topped with barbed wire.

Like many racially and environmentally unjust developments, For-mosa's Sunshine Project—named after the nearby Sunshine Bridge—has attempted to undermine not only Welcome's present Black population, but also its past. After Formosa's excavation team discovered the remains of at least four people buried in unmarked graves within a suspected cem-etery site, in 2019 it hired Alabama-based archaeology firm TerraXplora-tions, Inc., to conduct a more thorough investigation on its property. The firm's results were inconclusive, unable to identify the bodies as Black or white, and so Formosa has neither denied nor confirmed that the land they purchased holds the bodies of enslaved Africans or African Americans.

Yet a separate archaeology firm from Baton Rouge called Coastal Environments, Inc., conducted its own independent analysis. Working off a hunch that the plantation probably contained much more than the initial investigation yielded, members of the latter firm digitally ana-lyzed several nineteenth-century plantation maps. Just days after my visit to Welcome the firm revealed there could be up to seven cemeteries holding the remains of enslaved peoples and their descendants across the two former plantation sites now in Formosa's hands, based on their analysis. The neighborhood's residents, heeding the latest archaeological advice and their own roots in St. James, believe their ancestors are buried on the site of Formosa's planned plastic factory.

Residents of Welcome wondering what a new plastic plant would mean for their community can get an idea from Point Comfort, Texas, a small working-class port community about four hundred miles west. In Point Comfort, Formosa operates a plastic complex comparable in size and design to that proposed for St. James Parish. There, Formosa's biggest opponent has been Diane Wilson, a fourth-generation fisherwoman and retired shrimp boat captain who has spent more than thirty years documenting the company's pollution and challenging it in court. Nearly 40 percent of Point Comfort's residents are Latinx, Black, Asian, or biracial.[11]

Wilson's efforts include a 1994 lawsuit that saw Formosa agree to not discharge any plastic from its Texas facility—a promise the company ultimately failed to keep. She waded into the facility's outfall pipes along Cox Creek, where she scooped plastic pellets and powder into thousands of store-bought plastic bags. She used these samples as evidence in court to challenge the company, ultimately garnering a $50 million settlement in the largest Clean Water Act lawsuit ever filed by private individuals. US District Judge Kenneth Hoyt, who oversaw the case, declared Formosa a "serial offender" of pollution regulations and ordered the company to eliminate all plastic discharges or risk major fines. Hoyt also gave Wilson and her allies at environmental protection organization San Antonio Bay Estuarine Waterkeeper the green light to continue policing Formosa's pollution by monitoring waterways for plastic.

When Wilson heard about Formosa moving into Welcome, she began collaborating with St. James Parish residents to stop the company's latest development, testifying at legal hearings and providing the community with guidance and support based on her own experience.

To get to RISE's March 2020 meeting in Convent, she'd driven her pickup truck more than seven hours from her home in Seadrift, Texas. Formosa's Point Comfort complex sits less than twenty-five miles away.

"Our little community—with no support and no money—used legal aid. And when we did this, we won," Wilson explained to the crowd. "Your little community can do this, too. You can fight this."

Wilson dimmed the meeting hall lights and flipped through slides depicting the landscape of Point Comfort, Texas, a disturbing portent of what could be, in St. James, should Formosa persevere: white polyvinyl chloride powder blown over Formosa's grounds like a mad flurry of toxic snow, and the water and banks of Cox Creek coated with powder and white plastic pellets (nurdles) that sparkled in the sun "like little diamonds," Wilson pointed out. The crimson cape she wore flapped vigorously as she rapped the projector screen with an extended finger. "I tested some of the pellets I found in Lavaca Bay, where my family has been shrimping for generations. The pellets had mercury in them, and most likely many other chemicals. The shrimp and fish are eating these pellets. We are eating shrimp and fish from the bay at our own risk."

Anxious glances and expressions of disgust flashed from face to face.

"If you question Formosa, they'll use bribes to make the situation look better," Wilson said. "In Point Comfort, they've given away free watches and trips to the best resorts in Taiwan. That doesn't cancel out the fact that Formosa's neighbors live in a constant state of fear for their health and safety."

A woman with short blond hair and rose-patterned leggings named Harriet Livaudais Buckner waved a folded paper pamphlet in the air. "That's just like Dupont," she said. "I live in Gramercy. We get these brochures in the mail where Dupont talks about how it's paying for field trips at the local schools, giving back to our community by providing jobs. What they're doing is trying to brainwash us."

Heads nodded across the room.

"I went to the dermatologist recently and saw a mother crying because her five-year-old son's feet were encrusted with sores and eczema," Buckner said. "The mother said he can't leave home to go to school because she can't even get shoes on his feet—it's that bad. She said she lives in Reserve. Dupont is making synthetic rubber in her backyard; what are the chances her son's health problems are linked to that? And why is it that every one of us in St. James has had, or knows someone who has had, cancer?"

"If we don't stop this, soon kids will have to wear face masks and space apparatus just to go to the park," declared Stephanie Cooper, vice president of RISE St. James, teacher, and ordained minister at nearby Lutcher High School.

Though Cooper was referring to Formosa's expected pollution contribution to St. James's local environment, her comment would prove eerily prophetic: Again, it was early March 2020, just days before the Centers for Disease Control and Prevention would declare the coronavirus outbreak a global pandemic and much of the world—children included—would be required to don masks for protection against breathing in or spreading COVID-19. Formosa employees had begun staking out utility markers across the property, but in less than a week, the disease and seasonal creep of the Mississippi's spring floods would force their work in St. James to a standstill.

That month RISE had scheduled this and other meetings to strategize how best to continue applying legal pressure to stop FG. In early 2020, the company had obtained air pollution permits from the Louisiana Department of Environmental Quality (LDEQ)—an early step in setting up its operations. Shocked that the company's projected pollution levels passed muster, that February, RISE and an alliance of environmental organizations appealed LDEQ's approval of the air permits. The appeal asserted the pollution projections put forward by FG and approved by LDEQ were not grounded in science and vastly underestimated the additional chemical burden St. James residents would have to bear. They also charged that the department failed to address the concern that FG would be building its factory on top of the presumed burial grounds of enslaved African Americans.

"This blatant disparity in pollution burden in this community should be enough of a case for a lawsuit," said Kimberly Terrell, a biologist and the director of community outreach at Tulane University's Environmental Law Clinic in Louisiana, who also spoke at the March meeting. "Lawsuits cause major headaches for companies trying to get permission to build. It's usually easier to stop a factory that's planned rather than one that's already been built and is trying to renew or expand."

The appeal, in addition to another suit St. James residents filed against the US Army Corps of Engineers, was part of a clever tactic, Terrell reassured those St. James residents gathered before her, as it could require LDEQ to assess the factory's expected emissions and pollution control measures more stringently. That's useful, she added, because if St. James Parish residents can't stop Formosa's factory from being built, they at least want to ensure it is operating as cleanly as possible.

Following the appeal's filing, Janile Parks, Formosa's director of community and government relations in Louisiana, contested St. James residents' accusations that the Sunshine Project would cause irreparable harm to the community. She said the company's emissions models accurately project that pollution levels from the factory would fall within state and federal standards. Formosa, she said, "takes environmental and safety concerns about the Sunshine Project very seriously."

Formosa's track record tells a different story. The conglomerate, based in Taiwan, which runs more than thirty companies spanning petrochemical processing, electronics production, and biotechnology all over the world, has been accused or found guilty of polluting in every country in which it operates.

In 2018, scientists published a study finding that people living within six miles of a petrochemical and plastic complex in Yunlin County, which is located in west-central Taiwan, were more likely to be diagnosed with all kinds of cancers compared with those living farther afield. The complex in question is nearly 6,500 acres in size and contains sixty-four industrial plants, including oil refineries, power plants, and plastic production facilities. It's a major producer of polyvinyl chloride, or PVC, a type of plastic used to make pipes, wires, and car parts, among other items.[12]

Like St. James Parish in Louisiana, Yunlin County is rural and agricultural. The region's farmers have knit together a patchwork of manmade ponds teeming with farmed fish and thick flocks of domestic waterfowl, neatly planted rice paddies, and austere concrete buildings housing pigs and people. A tangle of towering smokestacks, flares, pipes, and other infrastructure is burned into the horizon, cordoned off by a dense brick-and-concrete fence topped with jagged shards of clear, green, and brown glass bottles. Explosions, fires, and other accidents have plagued the complex. In April 2019, a major explosion on the compound, in a plant that processes crude oil into chemicals used to make plastic, shattered glass of nearby structures, shook homes, and gave rise to a gas leak requiring the mandatory evacuation of more than ten thousand people in five nearby villages lying in the plant's shadow. This deadly petrochemical and plastic complex in Yunlin is owned and operated by Formosa.[13]

In 2020, residents of Yunlin publicly expressed solidarity with residents of St. James. "They don't want what happened to them to happen to other people," Xu Hui-ting, an environmental activist in Taiwan told *The World*.[14]

On June 19, 2020, the searing Louisiana sun bore down on a facemask-clad crowd descending from a collection of cars parked on the site of Welcome's former Buena Vista sugarcane plantation, on the planned site for the new Formosa Plastics plant. RISE and its partner organizations had gathered to commemorate Juneteenth, a holiday celebrating the end of slavery in the United States, while protesting the plant.

Racial tensions across America were especially high after the death of George Floyd in Minneapolis a few weeks before, another in a long list of Black people killed by police.

By eleven o'clock on Juneteenth morning, Sharon Lavigne's ranks had multiplied to dozens, despite the rising heat and the looming threat of the COVID-19 pandemic. Members of the St. James community, neighbors from elsewhere in Cancer Alley, and allies from outside the state of Louisiana settled into the tall grasses that grew along a towering chain-link fence topped with strings of barbed wire. From their vantage point, the crowd had full view of the probable slave burial ground. A Catholic priest clad in white, Father Vincent Dufresne, sprinkled holy water on the sacred earth. A few people carried a banner bearing a peaceful dove and the words "Honoring Our Ancestors." Staked into the grass below was another sign, one that took aim not at the past, but at the future, announcing: FORMOSA: YOU ARE NOT WELCOME HERE.

"Choosing to build this plant in a Black community sends a clear message," Lavigne said to the crowd. "They just want us to die off. . . . The intentional placement of industry here, and the poisoning of our community, reflects a racist society. But guess what: The air doesn't stop at the parish lines. If we stop industry here, we can reclaim St. James, but also other communities in Cancer Alley."

RISE St. James had filed a temporary restraining order against Formosa—which the company protested in court—just to stand legally at the fenceline of the cordoned-off cemetery site for one hour. Without the restraining order, Formosa could legally report any visitors as trespassers. At the last minute, RISE's restraining order was upheld. However, Lavigne later admitted, "I planned to visit the cemetery with or without official permission."

Less than a week prior to Juneteenth, Louisiana Governor John Bel Edwards vetoed House Bill 197, which would have criminalized protests on or near "critical infrastructure," including petrochemical complexes.

Proposed penalties included fines up to $5,000, plus three to fifteen years of imprisonment with hard labor. The bill's intentions were clear: to intimidate and bully potential protestors, including those like Sharon fighting against environmental racism, and to uphold the status quo.[15]

Just a few days before Juneteenth, a judge granted RISE and its guests permission to hold a one-hour ceremony at the Buena Vista cemetery site. After Formosa's security staff had checked all attendees' temperatures and recorded names and license plate numbers, they escorted the visitors to the site in their vehicles. Though the Sunshine Project was not yet standing, Formosa's factory—like all the other industrial complexes already constructed in Cancer Alley—had drawn a hard line between itself and local communities. This, though the plant's pollution and the products the company planned to manufacture would be impossible to contain.

The crowd prayed and sang and spoke in defiance of the continued injustices, environmental and otherwise, forced upon Black communities in America. As the single allotted hour on Formosa's property approached its end, Lavigne ushered her group toward the fenceline. In view of the gravesite, people threaded flowers through the chain link separating the living victims of slavery's legacy from the deceased who endured the cruelties of bondage firsthand. As soon as the last flower had been placed, Formosa's security squad began ushering the residents off the property in a slow procession of masked people and muddy vehicles.

CHAPTER 9

# Plastic and Our Warming World

Some may believe that, in a region already replete with chemicals, stopping Formosa—one plastic plant—would provide only trivial benefits to public health. In reality, such a victory would not only spare the residents of St. James from additional exposure to pollutants but would also be a win in the fight against climate change—another massive crisis we are now facing.

Scientists agree we must now wean ourselves off substances that contribute to climate change when extracted, processed, and burned—namely, oil shales, bitumens, tar sands, coal, petroleum, natural gas, and heavy oils—and we must stop continued industrial development.[1]

As Peggy Shepard, cofounder and executive director of WE ACT for Environmental Justice in New York and newly appointed cochair of the White House's first environmental justice advisory council told me: "It is essential to dismantle these institutions, because they are not going to concede power on their own. We must transition away from an economy based on fossil fuels, and that includes plastics. Our future depends on that."

And as we build up a new, renewable-powered world, "We need to ensure that it serves everyone—not just those currently in positions of power," Shepard added. Focusing on equity, rather than equality, she said, and prioritizing underserved communities' transition from fossil fuels and plastic production to renewable energy sources and materials is a good place to start. These efforts can help ensure the jobs, upgraded infrastructure, and other beneficial aspects of such a shift will serve communities affected by historical injustice, she said. "Decades of disinvestment driven by racism has put communities of color at a massive disadvantage. This is our opportunity to fix that."[2]

Our collective attention to climate change continues to grow as its deleterious effects—warmer air and seas, widespread wildfires, more intense and frequent storms, and species extinction, among them—become more apparent to us all. But humanity has collectively struggled to take the necessary step of shutting down the fossil fuel and plastic industries, because we have become entirely reliant on fossil fuels to navigate our hyperconnected, super-fast modern human society.

Recently, we got a glimpse of what leaning less heavily on fossil fuels could mean for us—and the industries we must disassemble: those dealing in oil, gas, petrochemicals, and plastic.

By April 2020, the coronavirus pandemic ground daily life to a halt for most people around the world: airports turned into ghost towns, empty gates amid shuttered duty-free shops; highways were devoid of cars; offices, restaurants, and entertainment venues closed and sent employees home. During the pandemic's peak, our carbon footprints were smaller than they'd been in a long time. Lockdowns, travel bans, business shutdowns, quarantines, and curfews forced people to stay local and make do. As a result, the world's collective carbon dioxide emissions dropped by 17 percent from 2019 levels.[3]

This is a not-insignificant number when you consider that the world's top climate scientists say global emissions must fall by at least 7.6 percent annually until 2030 in order for humanity to even slightly reduce the disastrous, rapidly accelerating consequences of climate change.[4]

Consequently, as the pandemic hit and demand for and values of oil and gas dropped precipitously, some smaller petrochemical companies were forced to shutter while a few larger companies issued temporary plant shutdowns and employee furloughs.[5] Meanwhile, residents of some of the world's largest cities—at least, those not living in wildfire zones—collectively reported that the air they breathed seemed cleaner than usual, even in some notoriously smoggy urban centers.[6]

As soon as regulations implemented to quell the pandemic were eased later that spring, global emissions began rising and air quality plummeted again, especially in industrial areas. By June 2020, lifting and uneven restrictions on travel and work pushed the world's greenhouse gas emissions back up to a measly 5 percent below 2019 levels, according to a report by the World Meteorological Organization.[7]

Although the slowdown in emissions caused by the COVID-19 pandemic was temporary, some experts, including climate activist Bill McKibben—author of *The End of Nature*, the first popular book written on global warming, published in 1989—have posited that petrochemical companies are finally losing some of their political and economic clout. "It's not a spent force by any means, but, even in the past few weeks, events have shown it to be waning where for a century and a half it has waxed," McKibben wrote in a May 2020 *New Yorker* article titled "Are We Past the Peak of Big Oil's Power?"[8] In his article, McKibben cites grassroots efforts protesting petrochemical development, university divestment campaigns, and the development of affordable renewable energy as major contributors to the fossil-fuel industry's downshift in power.

But the fossil-fuel corporations have a last-ditch plan to counteract diminishing demand: make more plastic. "It's no surprise to see that fossil fuel corporations have turned to plastics as a lifeline as climate change concerns reduce the demand for fuel," John Hocevar, Greenpeace's Ocean Campaigns director, told me.

Indeed, sensing a global shift in climate change policy, and reacting to new emissions agreements, Big Oil and Gas is banking on turning ancient carbon stocks—particularly shale gas—into plastic, instead of continuing to produce fossil fuels primarily to be burned for energy. In the US alone, major petrochemical companies like ExxonMobil, Saudi Aramco, and Shell have put more than $200 billion into several hundred natural-gas plastic and chemical facilities since 2010, according to the American Chemistry Council.[9]

Plastic production reached 311 million metric tons globally in 2014. That number is expected to double before 2030 and quadruple by 2050, according to analysts at the World Economic Forum.[10] Sales of petrochemicals, including those used to make plastic, regularly earn the world's top fossil-fuel dealers annual revenues in the tens of billions of dollars. Formosa Plastics landed at number six on the American Chemical Society's "Global Top 50 List" of petrochemical sales in 2019, netting $31.4 billion. Formosa is a family business built up by the late founder Wang Yung-ching, who died in 2008 with a personal net worth estimated at $6.8 billion.[11] Formosa Plastics is now chaired by Jason Lin, a longtime Formosa employee. The only companies to net more than Formosa in 2019 are Belgium's Ineos ($32.0 billion); Sabic, owned by the Saudi government ($34.4 billion); Dow, headquartered in the US ($43.0 billion); Sinopec, a Chinese company ($61.6 billion); and Germany's BASF ($66.4 billion). All are actively working to build up their plastic-production infrastructure.[12]

These super-wealthy corporations continue to target their development in underserved communities. In the US, plastic production is ramping up along the Louisiana and Texas Gulf Coast and in Cancer Alley, where so much petrochemical infrastructure already exists in communities of color. It's also expanding in the rural Ohio River Valley and Appalachia, where fracking wells brimming with natural gas are polluting thousands of low-income neighborhoods.

Plastic can be made from either oil or gas, and so are most of its additives. Plastic's main ingredients are pulled from freshly extracted fossil fuels in oil refineries and gas processing plants: naphtha, a crude oil–based substance; and ethane, a liquid natural gas. Naphtha and ethane are sent to so-called "cracker" plants, where immense heat, steam, and an absence of oxygen create olefin gases like ethylene, a key component of plastic bags; and propylene, the basis for much plastic packaging. Olefins are then sent down the pipeline to other plants for further processing, where they are turned into solid resins, or polymers—plastic. They emerge as pellets (nurdles) which are shipped to plastic manufacturing plants to be mixed with additives, melted down, and molded into products. Plastic complexes like Formosa's take care of many of these steps in one place, but sometimes plastic's petrochemical components change hands several times before emerging as a final product.

Extracting fossil fuels from the earth and turning them into plastic requires not only massive amounts of petrochemicals but also energy— and currently, this energy comes from burning more fossil fuels. According to the International Energy Agency, petrochemicals—including those used to make plastic—are set to become the biggest driver of growth for the global oil industry by 2050. The agency also expects petrochemicals to become a significant driver of gas industry growth: By 2030, global production of petrochemicals will require an additional 56 billion cubic meters of gas—approximately half of all of Canada's present level of natural gas consumption.[13] Greenhouse gas emissions linked to plastic production now hover around 900 million metric tons of carbon dioxide per year. That number is expected to surpass 1.3 billion metric tons, the equivalent annual carbon output of nearly three hundred coal-fired power plants, by 2030.[14] Those numbers exclude greenhouse gas emissions emitted during recycling and incineration, which also require energy, as well as landfills, which emit high levels of potent greenhouse gases, and from plastic itself.

Degrading plastic is a surprising additional source of greenhouse gases. Anywhere plastic is tossed by waves or wind and is exposed to sunlight—as it is when floating on the ocean's surface waters or when piled up on beaches—the material releases climate-warming gases into the atmosphere, including potent methane. Sarah-Jeanne Royer, currently a postdoctoral researcher at Scripps Institution of Oceanography at the University of California, San Diego, unintentionally discovered this phenomenon in the Pacific, which she revealed in a 2018 research paper.

At the time, Royer had set out to measure the greenhouse gases emitted by tiny organisms living in seawater. Like humans do, many of these organisms naturally release greenhouse gases as a byproduct of using oxygen to stay alive. She and a few colleagues scooped up some of these little ocean creatures with seawater in plastic bottles and began measuring the amount of methane they released. It was a lot—an improbably high amount. That's when her team quickly realized the plastic bottles they were using were also contributing climate-warming methane to their calculations.

From there, Royer turned her focus to plastic—specifically microplastic—and measured its methane emissions when exposed to sunlight in water and air. One of the most commonly produced and used plastics, low-density polyethylene (think: plastic bags), was the worst offender, releasing more climate-warming methane and ethylene than any other type of plastic she studied. Smaller, more weathered pieces of plastic released the most gas. Royer reasoned that microplastic particles' cracks and crevices give them a greater surface area from which gases can escape. In her research paper, she and her colleagues acknowledged the growing issue of plastic pollution, its inevitable transformation into microplastic, and its immortality. They closed with scientific evidence we can't afford to ignore: "The results from this study indicate that [greenhouse] gas production may continue indefinitely throughout the lifetime of plastics."[15] This is early research demonstrating that weathered plastic particles emit greenhouse gases. Scientists are not yet sure to what extent plastic is emitting these gases on a global scale.

It's clear: The more plastic people choose to put on the planet, the more forcefully humanity condemns itself to life on a dangerously warming planet.

Out on the Pacific Ocean, with its lack of people and infrastructure, it was easy to imagine what the ocean looked like before the Anthropocene: a pristine blue sea—filled to the gills with fish, whales, dolphins, turtles, rays, sharks, eels, porpoises, crustaceans, sponges, anemones, corals, plankton, jellies, and other creatures—that must have seemed to stretch on for eternity. But on land, imagining such a world requires a little more effort. As human population and reach has expanded, so has our species' contributions to a warming planet increasingly permeated by plastic waste. It's difficult to envision something different when living this way has been normalized by the industries disproportionately contributing to climate change and plastic pollution.

In the eighty-odd years since the first plastic items were manufactured and sold on a mass scale, the human population experienced its biggest growth surge yet. Almost every single person alive today uses plastic on a daily basis, most of which is designed for minutes or seconds of use before it no longer serves a designated purpose. So much plastic has come to surround us that some scholars insist that the mid-1900s heralded a new, unofficial, geological era within the Anthropocene, our present epoch of human impact on the planet. This period is called the "Plasticene," and it's defined by the deposition of a novel, synthetic planetary stratum across Earth's surface and seabed: petrochemical-based plastic.[16]

Ancient trash pits discovered in archeological digs around the planet reveal that human societies have long discarded artifacts they consider no longer useful. People are not unique in this regard; many animals have a tendency toward tidiness, especially those animals most commonly living in high-density communities. Central and South American leaf-cutter ants, for example, live in colonies of up to ten million individuals, occupying extensive underground nests in which they grow their own food—fungi cultivated from leaves cut from the forest. When ants or fungi perish, a designated group of leaf-cutters is employed to move the corpses or waste to a trash pile outside the nest, or to underground dumping chambers dug well below the ants' living quarters.[17] And while there's evidence that people and other animals are hardwired to separate themselves from their waste, overwhelmingly, living creatures are designed to waste as little as possible.

As recently as the early 1960s, the majority of people living across the United States and Europe who drank milk or soda had it delivered to their homes in glass bottles they could return for refill. People carried reusable cloth sacks to tote groceries home from shops and used tea towels to tuck away bread purchased at the baker. People in other parts of the world engaged in these practices even more recently. Writer and sociologist Rebecca Altman, an expert on plastic history, recalled learning a 1950s packaging magazine editor had told industry insiders, "The future of plastics is in the trash can."[18] Altman explained to me that the world had to be conditioned to carelessly consume. Prior to that time, she added, "it was not in the culture to use something once and throw it away."[19]

It seems scientists knew early on that one of the greatest dangers of plastic is its permanence. And while plastic's propensity for absorbing and leaching toxic chemicals has received much recent media coverage, this knowledge isn't entirely new. Some scientists revealed decades ago that if ingested in small amounts, "consumed particles of plastic could release sufficient amounts of PCB's to affect seabirds," as Stephen I. Rothstein, of the University of California, Santa Barbara, wrote in 1973.[20]

Field observations served as the basis for most of the earliest peer-reviewed research on how plastic was acting in the natural environment—especially in the ocean. In 1972, Aston University chemist Gerald Scott published a paper stating that plastic consumer packaging washing up on remote beaches was an ecological concern. In his paper, Scott discussed the slow speed at which plastic seemed to degrade in marine ecosystems and described a "need for the acceleration of this process" to avert sustained ecological damage.[21]

Also in 1972, Edward J. Carpenter, who is currently a biology professor at San Francisco State University, became the first person to publish warnings about what would come to be known as *microplastic*. That year, while completing a research stint at Woods Hole Oceanographic Institution, Carpenter published two historic papers: one that described "plastic particles" floating on the Sargasso Sea, and another that uncovered the existence of plastic nurdles—the same small, spherical pellets *Christianshavn*'s crew found inside the mahi-mahi caught in the eastern North Pacific Gyre—in waters and fish collected off the coast of southern New England.[22]

Throughout the decades following these early discoveries, just a few dozen papers on plastic pollution were published. It would take more than three decades from the time early researchers first announced they had detected small plastic particles in the oceans for the scientific term *microplastic* to appear in leading international journals. The publication of scientific research on microplastic is much more common today. A search of Google Scholar found 7,700 papers containing the word *microplastic* published in 2020 alone.[23]

While the first scientists to study plastic pollution described their concerns about the material's existence in nature and the bodies of wild animals, it took more than a decade following the publication of the earliest papers before the National Oceanic and Atmospheric

Administration (NOAA) called for a serious scientific discussion of the matter. In 1984, NOAA, which serves as the United States' primary science agency overseeing the ocean, hosted the first International Marine Debris Conference. The goal of the conference, according to former NOAA Alaska Fisheries Science Center deputy director Jim Coe, was to discuss whether or not marine debris, specifically lost and abandoned fishing gear, "was a problem worth people's attention."[24]

Scientists would swiftly conclude that it was. The conference's attendees agreed that plastic was accumulating in the oceans and recommended more research be done to better understand the problem. As a preventative measure, they also insisted immediate steps be taken to reduce plastic pollution discharged from ships. This compelled Congress to pay for the creation of the first iteration of NOAA's Marine Debris Program, the Marine Entanglement Research Program. Like today's Marine Debris Program, NOAA's earlier effort to address plastic pollution in marine ecosystems was designed to facilitate research, publicize data, and mitigate the problem.[25]

Despite this early concern, during the 1980s, plastic manufacturers remained focused on selling more plastic—which they accomplished by constantly reminding consumers about the utility of their plastic products, particularly plastic bags. Though former plastic industry executives—including Larry Thomas, who presided over the trade association Society of the Plastics Industry (today, Plastics Industry Association) at that time—would eventually reveal they were keenly aware plastic was ecologically harmful all along, they did not publicly acknowledge this problem. At the time, the plastic industry tried to show the opposite by emphasizing plastic's potential to be reused and recycled.[26]

In 1986, the industry-backed Plastic Grocery Sack Council told the Los Angeles Times: "Plastic bags can be reused in more than 17 different ways, including as a wrap for frozen foods, a jogger's wind breaker or a beach bag."[27] A few years earlier, the New York Times had published a story deliberating whether or not consumers would soon favor plastic bags over paper as grocery stores ramped up distribution of free plastic bags to shoppers. However, the story did not discuss the ecological problems a preference for plastic bags was likely to create.[28]

But even more than reusing, the plastic industry and major corporations pushed recycling as a salve to the waste issue their star material presented. When the 1970s wave of environmentalism prompted the public to vocalize their concerns about plastic waste, the plastic industry spent millions of dollars placing ads extolling the virtues of recycling plastic and developing plastic recycling systems in municipalities around the US. People across the nation, newly informed of the recycling imperative, were quick to participate. It didn't take too long to convince people it was OK—no, it was great—to throw things away. Plastic manufacturers even began stamping their products with numeric codes, which municipalities have long used to guide their recycling programs—telling people what they can, and cannot, recycle.[29]

Yet these numbers, called Resin Identification Codes, established in 1988 by the Society of the Plastics Industry (again, now the Plastics Industry Association) and now administered by ASTM International, are "not 'recycle codes,'" as ASTM International states on its website: "The Resin Identification Code is, though, an aid to recycling. The use of a Resin Identification Code on a manufactured plastic article does not imply that the article is recycled or that there are systems in place to effectively process the article for reclamation or re-use. The term 'recyclable' or other environmental claims shall not be placed in proximity to the Code." In other words, just because a plastic item possesses a resin code, this does not mean it is recyclable.[30]

"There's less plastic being recycled worldwide than the public is led to believe," Marcus Eriksen, 5 Gyres Institute cofounder and research director, has told me. Eriksen is a well-known contemporary documentarian of microplastic throughout the world's oceans and a campaigner against plastic and petrochemicals. He continued: "Many recycling systems fail because the precise sorting and cleaning they require is usually expensive and challenging to maintain. What's more, most plastics can be recycled just once or twice before losing their desirable qualities and become more prone to breaking apart quickly. It's cheaper and easier to bury plastic in landfills or burn it for energy. This fuels demand for more plastic, which is good for those who make and sell plastic."

And then of course there is all the plastic trash that's been labeled as "recycling" and shipped off to nations unequipped to actually recycle it, where it most often gets dumped in small, rural communities throughout Africa, Southeast Asia, Eastern Europe, and elsewhere. Governments have long swept the existence of garbage imports and exports under the rug—and the injustice this trash trade creates.

"Standing in between huge orange farms in Adana, Turkey, and over my head goes the most poisonous smelling blueish smoke, the kind that makes your head ache. The oranges are getting poisoned, the air is thick of dangerous toxins, and in front of me there's tons and tons and tons and tons of plastic waste from the UK!"

Sindy Yilmaz frequently airs grievances about garbage, like this one, to her online social media accounts. She reliably does so after her regular tours of fields rife with smoldering heaps of plastic waste dumped and set ablaze. They lie not too far from her home in the rural outskirts of Adana, Turkey. Accompanying her words are images of identifiable plastic objects within the towering piles of charred plastic.

Her images reveal much about the waste's origins: a tattered package of chicken bearing the red-white-and-blue British Union Jack, a faded green bag of pet food onto which English words were printed above the image of a crouching cat, a bent-up yellow UK license plate. . . . Yilmaz searches for languages and logos printed on plastic packaging, and other signs that may indicate provenance. Most of the trash she finds has not been generated in Turkey. To bring attention to her posts, Yilmaz often tags Turkish authorities and media outlets, in addition to those based in the countries from where the trash seems to have come. She claims authorities have dismissed her photos as fabricated and her claims as unfounded.

"The things people wouldn't do for money . . . such as bringing others' waste to their own homeland," Yilmaz lamented.

According to investigations by Sedat Gündoğdu, a microplastic researcher at Turkey's Çukurova University with whom Yilmaz shares her neighborhood finds, much imported plastic waste travels to communities in Adana through the Port of Mersin, on the Mediterranean Sea. After dumping plastic out of sight in the less-populated agricultural areas of Adana, recyclers often immolate their hauls—an attempt to burn evidence of their crimes. In an effort to hold Turkish waste importers accountable, Gündoğdu is presently working with the Global Alliance for Incinerator Alternatives (GAIA) to create an online map of illegal dumping sites that is accessible to the public. Gündoğdu said municipalities assume a position of "no transparency and no responsibility" when it comes to communication with Turkish neighborhoods targeted by dumping.

"The residents are really alone in this case," Gündoğdu said. "Adana is the hotspot for such illegal activities because of the illegal characters of the business. There are more than one hundred recycling facilities located in Adana. Most of the facilities are really far from any standards."

The devastating effects of humanity's plastic pileup are particularly acute in Turkey and other cash-strapped nations long treated as dumping grounds for others' trash. While the Basel Convention was signed in 1989 to prevent movement of hazardous waste across borders and updated in 2019 with amendments limiting shipments of plastic waste, for decades nations have been exploiting a loophole in its wording authorizing overseas shipment of waste so long as it is labeled "recycling." These agreements are consensual; plenty of importers and exporters alike have made billions off of trading waste over the past few decades. Their secret is to keep costs low. All they have to do is recycle as little as possible, if at all. More often, recyclers—sometimes aided by authorities—simply dump and burn imported plastic in someone else's neighborhood. What is exported to other nations actually tends to be the least recyclable, most problematic, types of plastic: that of the one-time-use variety, such as single-serving sachets, food wrappers, and plastic bags.[31]

Like neighborhoods bordering fossil-fuel extraction sites and the industrial barbed-wire fencelines behind which plastic and its petro-chemical ingredients are made, neighborhoods drowning in imported plastic tend to be those most often underserved by society. And they too are constantly living on the precipice of the next emergency. Festering piles of untouched plastic spark frequent fires at antiquated, neglected recycling plants. Landfills leach toxins into soil and waterways and send climate-warming gases into the sky. Unauthorized dumps set ablaze can burn out of control, igniting forest fires and threatening homes. Some importers send plastic to incineration plants. These facilities, generally overwhelmed and aging, are prone to frequent fires, explosions, and other accidents bound to happen when flames are brought to one of the most flammable substances on the planet: plastic.[32]

When plastic burns, it releases dioxins, phthalates, furans, mercury, PCBs—chemicals known to increase heart disease and cancer risks, worsen breathing problems, harm the nervous system, and reduce fertility, in addition to causing headaches, rashes, and nausea. Burned plastic also releases soot, worsening deadly particulate air pollution, and greenhouse gases, contributing to climate change. At incineration plants, the remnants of what is burned—gray, powdery fly ash—is typically landfilled or poured into poorly contained holding ponds tainted by toxic lead, copper, and cadmium. Out at dump sites, plastic's chemical-laden remains spread and seep into soils and groundwater. Unburned plastic breaks up into microplastic. People routinely exposed to toxic plastic smoke and particles suffer from constant discomfort and chronic health problems.[33]

Also suffering disproportionally from health issues linked to plastic and other toxic wastes are the millions who make wages from selling recyclable items pulled from dumps, bins, and the natural environment. For many reclaimers, waste picking is a primary source of income, though it is also highly hazardous. Wading through piles of plastic and other waste invites serious injuries, infections, and respiratory problems, which tend to go untreated. Reclaimers generally have scant access to health care and other types of aid, despite providing an important service, clearing trash out of their communities to give it another life. In some regions, waste picking is the only type of reliable recycling system available.[34]

Forty-five percent of all nations' exported plastic "recyclables" have been shipped to China—which itself manufactures more plastic than any other country in the world—since 1992.[35] The global flow of plastic to China, the world's biggest plastic importer, came to a grinding halt in 2018 when it closed its borders to most all detritus sent from other nations, citing environmental concerns.[36] This new policy, referred to as the National Sword, sent exporting nations scrambling to reroute their recycling shipments elsewhere. Mainly, the surplus plastic not sent to China was delivered to Malaysia.

Turkey, too, upped the ante as an importer, accepting plastic from abroad at record levels: about 48,500 metric tons of plastic each month throughout 2019, according to Greenpeace Turkey. Most of this trash was delivered by the UK, Italy, Belgium, Germany, and France.[37] Meanwhile, plastic factories located throughout Turkey churn out increasing amounts of freshly made plastic, most destined for a future as packaging that will readily be discarded—and probably never recycled.

Despite all the plastic around, widespread public awareness of Turkey's import and illegal dumping problem is low, according to Gündoğdu. In fact, awareness of the pervasiveness and magnitude of the plastic crisis around the world falls far short of where it might otherwise have reached, had the history of plastic unfolded differently.

According to contemporary plastic pollution scientists, industry campaigns designed to keep people in the dark have contributed significantly to a general lack of public knowledge about the plastic crisis. This may help explain why efforts to address plastic pollution and its associated injustices—whether they are stopping petrochemical development or illegal plastic imports and dumping—have been so excruciatingly incremental. From day one, the plastic industry has been working to carefully curate its public image.

A long-used tactic still employed by the plastic industry has been to extoll the virtues of recycling as a solution to plastic pollution. As Eriksen has explained to me, it's "a way to deflect attention and responsibility for product design away from industry, and has been very effective. Industry has aggressively defended themselves, manipulating public perception, and attacking scientists perceived as a threat."

"For both papers in *Science*, the Society of the Plastics Industry sent a representative (twice) to Woods Hole, basically to intimidate me," stated Edward J. Carpenter, the early plastic researcher. "I was not given tenure at Woods Hole Oceanographic Institution, and I think the plastic papers hurt my career there."

The Plastics Industry Association refused to confirm or deny Edward's claims when reached for comment.

Similar information-suppressing strategies have been applied by industries to hide the relationship between tobacco use and cancer, and fossil-fuel use and climate change. It seems many corporations stick to the same set of strategies when attempting to obstruct legislation and shrug off liability for issues their products create.[38] Following what seem to be the same predictable plays as other deceptive businesses, the plastic industry has harassed scientists who share inconvenient results or views, diverted attention from scientific recommendations (chiefly, to reduce plastic use), and made strong attempts to block policies they find unfavorable (such as banning or restricting plastic use), among other tactics.

In 1993, Denmark passed the world's first legislation restricting distribution of lightweight plastic bags and taxing heavier plastic bags, the kind of thicker polyethylene that resists tearing and so can be used several times before wearing out. Today Danes use an average of four single-use plastic bags and seventy of the thicker multiple-use plastic bags a year. In the US, where the vast majority of states and municipalities do not regulate distribution of plastic bags, people use an average of one lightweight single-use plastic bag—the kind most frequently found littered or escaped from rubbish receptacles, blowing around beaches, agricultural fields, and roadsides—per *day*.[39]

The industries making and using plastic, and supplying fossil fuels, have actively fought such legislation aimed at curbing plastic manufacturing and use, especially in the US—the world's biggest generator of plastic waste.[40] Starting in the 1970s, these industries began forming deceptively named pro-plastic campaign groups such as Keep America Beautiful and American Progressive Bag Alliance, which have lobbied against regulation that would curb plastic production and use, preventing plastic pollution. Meanwhile, the plastic industry has continued to sell its products while pushing the present broken recycling system as the best method to reduce waste and litter.

"The public did not get adequate information, or the right information, early enough to act," said Eriksen. "Industry has been very effective at controlling the public narrative."[41]

Although its public messaging suggests otherwise, the plastic industry hasn't been able to control the spread of plastic nor the people who bear witness along its trail of destruction. Every voice speaking out helps grow a wave of awareness. And some voices have managed to capture the world's attention, playing a major role in shifting our relationship to plastic. One of those voices belongs to Charlie Moore, who has gone on to share his story across mainstream media as discoverer of the Great Pacific Garbage Patch.

In 1994, Moore, a sailor and scientist, founded the Algalita Marine Research and Education Foundation to survey and address humanity's deleterious impacts on and varied exploitations of his home waters off southern California. A year later, he acquired a fifty-foot, twenty-five-ton aluminum-hull catamaran in Hobart, Tasmania, and launched her as his organization's flagship research vessel, *Alguita*. As Moore and his small crew sailed the ship back to a slip by his home in Long Beach, California, stormy weather managed to rip off her mast. Fortunately, *Alguita* was rescued, locked up in a container, and shipped to Long Beach for repairs. Two years later, in 1997, Moore had the nerve to enter his newly rerigged research vessel in the "Transpac," a world-class yacht race from Los Angeles to Honolulu, to test its seaworthiness. Incredibly, *Alguita* proved herself fit for the journey of more than two thousand nautical miles, traversing the route from Los Angeles to Honolulu and avoiding the windless gyre swirling farther north.

On the way back to California, as the other sailors headed back along the speedy oceanic racecourse, Moore decided to take the scenic route: With extra fuel and two engines his only security from being marooned in the doldrums, he diverted *Alguita* through the gyre. Due to its lack of wind, the gyre was—and continues to be—a place rarely visited by sailors. And yet, despite being so far away from any other people, Moore and his crew spotted a disturbing number of plastic items floating on the sea.

"I was shocked to find there the detritus of civilization," the sea captain reflected. "Plastic had snuck away to this pristine place in the middle of nowhere, and at that time, it seemed no one knew about it but us."

He called this place the "Great Pacific Garbage Patch." Soon, Moore was traveling the world sharing his story, speaking widely about plastic pollution in the media.[42]

Altman, the plastic history expert, has suggested that, along with industry's push to silence research and shape consumer ideas of waste, the media has also played a part in plastic pollution's longtime obscurity and recent rise to global consciousness. She noted that this rise to prominence seems linked to a combination of plastic pollution worsening and the nature of media changing over time. Today, social media gives anyone, anywhere the ability to share what they think is being overlooked, and they're more disposed to post about visibly compelling big stories—like plastic pollution. Just think of the Great Pacific Garbage Patch; by this point, it should be old news, yet it has still received a surge of media attention in the past decade.[43]

"Culturally, we focus on environmental problems of a spectacular nature, the kind of havoc that happens in a bewildering instant," Altman said. "It's hard to see the slow-moving disasters or tragedies that happen over time—the *drip, drip, drip*—until it's of a disastrous proportion."

Edward J. Carpenter has agreed with this perspective, noting the long lag between the scientific discovery of plastic pollution in the ocean and publicity about the problem. "I believe that the Captain Moore TED Talk on the Great Pacific Garbage Patch, plus Marcus Eriksen of 5 Gyres, plus a video on dying albatrosses at Midway Island, plus the graphic video of the sea turtle with the plastic straw up its nose began to finally wake up the public," he told me in 2019.[44]

Carpenter himself pointed it out nearly five decades ago: The more plastic we make and use, the more will end up in the natural environment. As he wrote in 1972, "Increasing production of plastics, combined with present waste-disposal practices, will undoubtedly lead to increases in the concentration of these particles."[45]

If we'd listened back then, life on Earth could have looked very different today.

(*From left*) Henrik Beha Pedersen, Malene Møhl, and Lisbeth Engbo of Plastic Change; with Nohea Ka'awa and Megan Lamson of Hawai'i Wildlife Fund, at a cleanup on Kamilo Beach, Hawai'i, in December 2016. Photo by Erica Cirino.

PART IV

# Solutions

# CHAPTER 10
# Cleaning It Up

"We return week after week doing cleanups and we know the plastic will come back. The thing we need to do is bring attention to the problem, help people understand what's happening out here is related to the way they live at home," Megan Lamson explained as she stepped out of her rusty Tahoe onto the beach, scruffy brown dogs spilling out after her. Lamson knelt down and scooped up what seemed like coarse sand. Up close, one could recognize its true composition: lightly colored, weathered bits of microplastic of various shapes, sizes, and textures—round, squared, small, tiny, filmy, hard, soft, sharp—mixed with little grains of light sand, flecks of white coral, broken shells, and bits of black volcanic rock.

Strewn across the beach's plastic-and-sand coating was an impossibly vast and diverse collection of plastic items. Some were weathered to the point of being unidentifiable, worn and fragmented by wind, waves, and sunlight; though many were indeed still intact and recognizable, despite having spent considerable time in the ocean as indicated by layers of algae or smatterings of barnacles, and distinct styles marking specific moments in time: flip-flops, clothes hangers, disposable hair combs, children's beach toys, fish crates, a rusty refrigerator, threadbare tires, straws, bottle caps, buoys, toothbrushes, lighters, golf balls, and other items people have produced over the past eighty years and counting.

"Well," Lamson declared after her quick scan of the beach. "It doesn't look too dirty today."

While newcomers would perceive Kamilo as absolutely saturated with plastic—indeed, it has gained the reputation as one of the world's most polluted beaches—to Lamson, president and program director at Hawai'i Wildlife Fund, this was just another day at the beach. She and the other staff and volunteers working to improve the health of Hawai'i's environment have made near-weekly trips to remove plastic debris from Kamilo Beach, located on the southeastern end of the Big Island of Hawai'i, for many years.

With her surfer's prowess for reading waves, Lamson had learned to predict and evaluate the beach's plastic load with supreme accuracy just by looking at the weather. There'd been a few storms near the south point of the Big Island recently, she pointed out. This meant rowdy waves had washed some of the plastic that had found its way onto Kamilo's shore back out to sea. So in fact, less plastic was present than usual that day. Yet there was still much work to do.

I met Lamson at the beginning of December 2016, with a few members of *Christianshavn*'s crew and Plastic Change's Danish media team, to assist with and document a cleanup. After sailing from Los Angeles to Honolulu, the sailors took the forty-five-minute flight from Honolulu to Kona. We were accompanied by Lamson's then-fiancé Patrick Leatherman and Nohea Ka'awa, a Native Hawaiian and practitioner of her culture who has worked in community outreach and education for the state and nonprofits including Hawai'i Wildlife Fund. Lamson, Leatherman, and Ka'awa wasted no time pulling large orange construction buckets, worn sacks, and salt-stiffened cotton gardening gloves from the backs of the trucks. As our cleanup team prepared, the dogs ran circles around us, tearing up stray coconut husks and dipping into shallow waters gleaming beneath an overcast sky. When they emerged, many flecks of light-colored microplastic clung to their soaked fur.

Although we're on track to expel a staggering fifty-three million metric tons of plastic into waterways in the year 2030, if you're not in an area of concentrated plastic like the eastern North Pacific Gyre, it can be hard to understand the true magnitude of plastic in the ocean—the place where much of our plastic ends up.[1] In places like Kamilo Beach and other hyper-polluted beaches and ports, however, the plastic pollution crisis is a constant and visible threat.

Out at sea, wind systems, Earth's rotation, and the location of landmasses form gyres, eddies, tides, and currents that readily transfer heaps of plastic that have entered the ocean back to shore. Some of the planet's beaches become more blighted than others. That depends on their location: Sheltered beaches are more likely to avoid the brunt of trash tides (though that's no guarantee against people littering shorelines with plastic). Island chains out at sea, unprotected by continental landmasses, are a prime landing spot for plastic set adrift, and tend to bear the brunt of plastic spilled from shipping containers lost at sea. Some of the worst-polluted beaches are also the most remote. In the Pacific, the North Equatorial Current pushes water west along the upper bounds of the equator, helping spin plastic from the Garbage Patch to an undeveloped area on the southeastern coast of Hawai'i's Big Island: Kamilo Beach.

The problem may seem overwhelming, but at Kamilo, Lamson and a dedicated group of plastic collectors use the pollution as a way to spread awareness of the wide range of issues linked to plastic—and prevent the trash that has collected there from washing back out to sea.

More than seventeen metric tons of debris washes up on Kamilo Beach every year, and Hawai'i Wildlife Fund reports that it pulls about twelve to sixteen metric tons of that debris off a seventeen-kilometer stretch of its coastline, most of which—about 85 percent—"is plastic stuff," as Lamson explained. "The next biggest category is broken glass, but I tell volunteers not to pick it up, because it's dangerous and, unlike plastic, eventually degrades."

Malene reached down to pull the day's first piece of plastic—a child's red plastic trowel—from the beach. As she moved to drop it in a bucket, Lamson gently pressed Malene's arm, holding it back.

"Wait," she said. "We can't touch the beach yet. It's a sacred space, so we must ask for permission to enter and do our work."

Malene nodded, knelt down, and gingerly placed the trowel back on the sand.

Megan turned to the group. "Repeat after me," she began, and proceeded to recite a short chant in Hawaiian, called an *oli kāhea*. The oli kāhea is a part of traditional Hawaiian culture—essentially, a verbal request people are expected to make prior to entering others' spaces. It is customary in Hawai'i to ask for the sea's permission to allow us to tread on—and, in our case remove plastic from—her shores.

We repeated the chant and then paused. Ka'awa followed up with another chant, the *oli komo*, the response to oli kāhea. Our required display of respect for the ocean complete, we descended on the beach. Henrik was immediately pulled to the water's edge, where he submerged a hand, scooping out a heaping handful of microplastic. Seawater infused with a swirl of colorful microplastic surged toward his sneakers, the gentle shore break on Kamilo more closely resembling the contents of a vigorously shaken snow globe than ocean waves.

We filled our orange construction buckets with debris and then emptied the buckets into the big burlap sacks, thick clear plastic contractor bags, and jumbo-sized dog kibble bags Lamson had brought to the beach. With a hint of pride detectable in her voice, Lamson noted that they've been reusing the same collection bags for decades.

Lisbeth Engbo, then working in communications for Plastic Change, plucked hundreds of plastic bottle caps from the sand, carrying the colorful objects in a large plastic bowl that had also washed up on the beach, as if filling a basket with fresh-picked fruit. Malene sat in the sand, attempting to separate tiny microplastic bits from often equally tiny grains of sand with a fine sieve. After more than an hour's effort in the same small patch of sand and only a few handfuls in the bottom of her bucket to prove for it, Malene rolled her eyes and said, "Now I need to get up and fill up a big bucket with larger pieces of plastic, or I'll be unsatisfyingly pushing around microplastic all day." It would take "close to forever" to attempt to pick up every piece of plastic off the face of the planet, she concluded. "And even then, we'd probably miss the majority of it."

With growing attention on the plastic crisis, many people are trying to clean up the plastic already released into the environment. Much of the work is being done on beaches, like Kamilo, where much trash tends to collect. In the past fifty years, beach cleanups have become mainstream and continue to play an important role in spreading awareness of the plastic crisis, among other benefits. One major contribution to this effort was Ocean Conservancy's establishment of an annual global beach cleanup day, called the International Coastal Cleanup, in 1986. To date, more than 15 million people from more than 150 countries have removed more than 300 million pounds of trash from beaches worldwide as part of the International Coastal Cleanup, which is held each September.[2]

Some individuals and organizations have turned to technology in an attempt to scale up their cleaning efforts. In 2018, Boyan Slat, a Dutchman, then twenty-four years old, directed the delivery of a $20 million, two-hundred-meter-long boom he had designed with his nonprofit The Ocean Cleanup to catch microplastic and plastic items in the eastern North Pacific Gyre. The mainstream media and other innovators widely praised Slat's invention (which he called "System 001") as it was towed from California to the gyre. However, some scientists assessed Slat's costly gyre cleanup attempt as "a waste of effort."[3]

Ultimately, Slat's plastic-catching boom—itself made of plastic—began disassembling soon after it was deployed. As plastic does. The broken boom was towed back to shore for upgrades. Slat and his team managed to redeploy an upgraded version in 2019, and in less than fifty days it collected a little more than ninety metric tons of plastic debris, which The Ocean Cleanup says it has manufactured into luxury sunglasses. That's not enough plastic to keep pace with Slat's goal of removing 50 percent of the plastic presently in the gyre in five years, and 90 percent of all plastic in the ocean by 2040.[4]

Despite these setbacks, Slat has continued his work with The Ocean Cleanup and is presently at work on a new fleet of devices that could more efficiently catch plastic in the ocean. However, today he and his team have shifted much of their focus toward catching plastic closer to shore, before it flows out of rivers—which experts agree is a more efficient and feasible way to intercept debris destined for the sea.

With similar upstream efficacy, the "Seabin," a bucket-shaped device initially designed for installation in marinas, sits just below the waterline against a piling or dock and gently pulls floating plastic into a secure container. Seabin's inventors, boatbuilders Andrew Turton and Pete Ceglinski, founded the Seabin Project in 2015 after discussing whether the seas might be cleaner if floating trash bins existed. Today, hundreds of Seabins are now in use in marinas around Europe, North and Central America, the Middle East, New Zealand, Tasmania, Australia, and Asia. Turton and Ceglinski estimate the devices have the potential to capture about two to three pounds of trash per day, totaling about a half a ton of debris per Seabin per year. The team is continually working on upgrades but has found that their passive trash-collecting bins seem to do a decent job of reducing pollution in nearshore waters. Despite the

Seabin's ability to locally reduce pollution in marinas, the Seabin Project acknowledges: "The Seabins are not the solution to plastic pollution. We believe that a real solution lies in education, science, and systemic change."[5]

In Maryland, John Kellett, former director of the Baltimore Maritime Museum, developed a "Trash Wheel" that lifts plastic and other debris from the mouths of creeks, falls, rivers, and streams before the trash flows to the sea. Part old-fashioned waterwheel, part cartoon character (the fifty-foot-long machines are decorated with large googly eyes and have their own social media accounts), Kellett's first solar- and hydro-powered trash-conveyor wheel was installed in 2014 to intercept trash before it reached Baltimore Harbor. He devised his wheel after observing thick streams of litter flowing from Jones Falls into Baltimore's Inner Harbor every time it rained—this was trash flushed from streets into storm drains that led to the falls.

There are now four trash wheels installed around Baltimore. Depending on the day, and how much it has rained, the trash wheels may pull out several tons of trash, primarily plastic, in a single day. After strong rains in April 2015, Kellett's first machine, nicknamed Mr. Trash Wheel, which sits at the mouth of Jones Falls, scooped a whopping nineteen tons of trash from the water in twenty-four hours.[6] In the nearby Anacostia River, water and sewer utility DC Water has deployed two fifty-foot-long trash-skimming boats, named Flotsam and Jetsam, which also use conveyor belts to scoop up plastic and other waste. Others are working to install similar trash-catching devices, including booms and grates, in waterways around the world to stop plastic from reaching the ocean.

While the world's coastlines and freshwater systems are massive, they're nowhere near as vast and complex as the ocean. While locating and pulling plastic particles and products from the sea is proving tricky and time consuming, a task too tough for even the most advanced technologies to carry out, plastic accumulation zones on beaches and in waters upstream make trash more accessible to people who seek to clean it up. There we can manage to prevent a small but important fraction of plastic waste from entering the sea, or remove what has washed in from the ocean, using technology, and with our own physical labor—plucking up plastic piece by piece.

Cleanups do more than divert plastic from the sea. Wherever trash is collected, it can be accounted for, helping us better understand the composition and fate of plastic marine debris, hopefully leading to solutions that prevent more trash from getting into the natural environment. And yet, the fastest way to stop plastic from flowing into the ocean is to stop it at the source: Use less or, ideally, no plastic. "You wouldn't just mop up water off your floor if your bathtub were overflowing," Malene once told me. "You'd turn off the tap."

To address the plastic crisis, we must stay focused on the ultimate goal of curbing plastic production. And while beach cleanups don't do this directly, it's something they are helping to communicate. "It's mainly all the single-use packaging we constantly buy and discard that is causing this problem," Lamson said. "So many people who clean beaches, Kamilo especially, tell me that they've gone on to reduce their use of plastic. This is a pretty shocking thing for most people to see."

And it is shocking. Spending hours stooped over a beach to pluck sunbaked plastic particles and objects from its surface, and recognizing so many of the plastic items you have been instructed to throw away your entire life, gives a person a lot to think about.

Malene joined Sofie at the beach's vegetated perimeter to collect tattered plastic bags—now just eerie polyethylene ghosts—that had blown from sea to scrub. Henrik moved fast across the vast beach, surveying the scene. He scaled a dark rock outcropping covered by an array of cracked plastic fish crates emblazoned with the elegant markings of

Japanese characters, which he dragged into a messy pile. I walked with Lamson down the shoreline, searching for birds, Hawaiian monk seals, hermit crabs, sea turtles, and other animals—ready to free any creatures unlucky enough to get trapped in the beach's plastic obstacle course.

Another major benefit of beach cleanups, said Lamson, who is also a marine biologist, is their usefulness in removing plastic that might harm wildlife. "Every piece of plastic picked up could potentially save a life," she noted. But at that time, she had noticed a growing catastrophe affecting Hawai'i's turtles—particularly rare hawksbills—that was too large to easily clean up.

As hinted at in their name, hawksbills have a decidedly avian-inspired aura: These reptiles wear shells mottled by feathery orange, brown, red, and gold strokes, and they have down-curving beaks reminiscent of their namesake raptor. One of three species of sea turtle living in Hawaiian waters, hawksbills live on coastal coral reefs from which they pry sponges, anemones, squid, and shrimp using their hooked beaks. Unsurprisingly, as the effects of climate change, development, and pollution intensify and continue to destroy corals, life as a reef-dependent hawksbill has become much harder to hack. What's more, hawksbills' distinctive painterly shells have long made them a highly sought-after target for poaching and trade, activities that further erode their numbers.

On top of all these challenges to hawksbill survival, plastic is pushing them even closer to the edge of extinction. Hawksbills, like all other sea turtle species, are regularly ingesting plastic pieces big and small. Globally, more than 50 percent of all sea turtles are thought to have consumed plastic at some point in their lives, according to Australian sea turtle expert Dr. Qamar Schuyler from the University of Queensland.[7] Schuyler and her colleagues have also found that eating even a small amount of plastic can be deadly: A sea turtle that ingests a single piece of plastic has a one in five chance of dying from it. The more plastic a turtle eats, the worse off they are. Scientists say a sea turtle who has eaten, for example, fourteen pieces of plastic has a 50 percent chance of dropping dead in the very near future.[8]

Lamson gestured at the microplastic-sand-coral-shell-rock mixture that blanketed the beach. Turtles are not only eating plastic, she explained. "Then there's all the microplastic that's mixed with sand on our beaches which may also be killing our hawksbills by messing with their eggs. A warm beach can turn a whole clutch of eggs female—then whom will the female turtles mate with, if anyone at all?"

Embryos inside sea turtle eggs acquire a male or female sex based on the temperature of the surrounding sand. Warmer sands usually nurture the development of a greater number of female turtles, while cooler sands tend to bring more males into being. As plastic particles compete for space with sand on beaches, shoreline temperatures soar. Microplastic absorbs heat, raising the temperature of surrounding sands and causing a greater number of females to hatch, a dire trend that will only worsen as the warming effects of climate change rapidly accelerate. Lamson explained that a higher ratio of females to males could reduce the overall reproductive success of hawksbills—a highly vulnerable species to begin with, due to its being targeted by poachers and dependence on disappearing reefs. Scientists have already begun noticing extremely feminized sea turtle populations in some parts of the world, including the Great Barrier Reef and in Cyprus, Greece.[9]

Pregnant female hawksbills have been known to "false crawl" toward highly polluted beaches, including Kamilo—hauling up onto the sand briefly before retreating and returning to the water without laying any eggs—apparently repelled by the strange plastic obstacles they find scattered across the shore. And if females do manage to navigate the piles of plastic well enough to lay eggs, Lamson said, their hatchlings are often impeded by plastic debris, which can injure or trap the inch-long youngsters as they attempt to make their way from land to the ocean and begin their lives.

"I expect we'll keep finding fewer hawksbills around our islands, and of those we do find, there will be more females than males," said Lamson, who has organized a hawksbill sea turtle recovery project through Hawai'i Wildlife Fund.

The project involves recruiting volunteers to monitor beaches for hawksbills and keeping their nests safe from an endless accumulation of plastic debris, trampling human feet, and hungry dogs—among other dangers—with beach cleanups, monitoring, educational campaigns, and the installation of fencing around nests during hawksbills' forty-seven- to seventy-five-day incubation period.

"And I mean, these are my predictions if eggs hatch at all," Lamson added. "We're seeing a lot of eggs that simply don't open up. Could it be the chemicals from the microplastic sand seeping into them and killing the baby turtles? We don't know yet."

I spotted Ka'awa sitting in the sand, sifting for microplastic, picking up the tedious task where Malene had left off. Ka'awa is considered a *Kama'aina*, "a child or person of the land," because she was born and raised on Hawai'i—more specifically in Ka'ū, where her family has lived for generations. I sat down next to her and asked how did she, as a Native Hawaiian, feel about the plastic strewn across their shores?

"What makes me sad is that many of my people—and most other people—don't realize that *they* are causing this problem. The environment means so much to us: The Earth provides for us, it's our ancestors giving us life. We're acting like an invasive species that is disconnected from our natural world, our lifeblood," she explained.

Ka'awa was quick to point out that, unlike many other cultures, which paint human existence in a realm outside nature, Hawaiian beliefs integrate humanity as a part of nature. Traditional Hawaiian culture encourages people to respect the nonhuman world in order to achieve a balanced life of morality, a state of being Hawaiians refer to as *pono*.[10]

For example, many Native Hawaiians have believed that deceased people are reborn after death as stones, trees, stars, clouds, and local nonhuman animals—such as albatrosses, referred to locally as *mōlī*, and green
sea turtles, or *honu*. These nonhuman embodiments of the deceased
are thought of as guardian ancestors, or *ʻaumākua*, who are thought to
provide strength, guidance, and help to their family and friends who are
still alive.

In Hawaiian beliefs, gods and goddesses are also likely to have natural embodiments.[11] Native Hawaiians' creator figure is a goddess of fire
and volcanoes, named Pele, whose molten spirit formed the Hawaiian
Islands. She is thought to exist in every spit, spill, and splatter of steam,
rock, lava, and ash that rises from the Big Island, especially the molten fires emanating from the crater of Halemaʻumaʻu, part of Mount
Kīlauea in Hawaiʻi Volcanoes National Park—one of the world's most
active volcanoes.[12]

Because nature is deeply sacred to Hawaiians, desecrating it—such
as removing a hunk of lava rock from a beach—is understood as an act
of ill will, betrayal, or evil. Kaʻawa explained that today many Native
Hawaiians—particularly younger generations—are less likely to engage
in traditional cultural practices and more likely to be influenced by glo-
balization and capitalism, and so seem increasingly disconnected from
their roots in nature. Over the years Kaʻawa said she's noticed an influx
of trash generated by tourists and locals alike. She only expects this
trend to intensify into the future as social media and the ad-heavy internet pushes people to consume more.

"That is, unless we begin to remind people, especially young people,
that we are not living in balance and that is harming the planet," said
Kaʻawa. "All people must recognize that without a healthy Earth, we
cannot survive."

As we filled the last of the bags and buckets with plastic debris, a
hulking black Ford F-250, rigged into a homemade dump truck with
metal and wooden planks, rumbled down the path. The truck towed a

large trailer painted in green camouflage, similarly outfitted with wooden supports—both apparently built to hold heaps of plastic debris. A man with gray hair tied into a short ponytail emerged from the pickup-turned-dump-truck and introduced himself as Bill Gilmartin, a cofounder and vice president of Hawai'i Wildlife Fund. Ka'awa tossed bags up to Henrik, who stood in the bed of Gilmartin's truck, while Leatherman emptied more bags into the trailer. Lamson tucked a small bag of microplastic into her truck, which she said would be delivered to a local artist, one of many around the world recognizing plastic debris as an art medium.

After the day's outing, Gilmartin said he would tip the plastic debris loaded onto his truck and trailer into a nearby dump. "It's not the best solution, but it's the best we can do right now," Gilmartin admitted, well aware of the constant and massive leak of plastic from modern waste management systems into the environment.

What to do with plastic once we pick it up remains a major challenge. The fate of most plastic collected by cleanup devices is identical to that of all plastic at the end of its life—it's dumped, incinerated, or shipped off. Some will inevitably make its way back out to sea. A much smaller amount of plastic collected during cleanups and caught by cleanup devices is recycled. Plastic debris pulled from the natural environment is even more challenging to recycle than used plastic items shipped straight to a facility. While plastic litter is technically a free resource anyone can tap, for most purposes it must be cleaned and processed and even then may not be suitable for reuse; as we know, weathering and exposure to heat, water, and physical forces reduces the strength of plastic and can introduce toxic chemicals, bacteria, or viruses.

Despite these obstacles, a growing number of people are using plastic collected during cleanups or by cleanup devices to make products sold to engage people in conversations about plastic—and spread awareness around why it's a problem and what to do about it. One item that qualifies is The Ocean Cleanup's aforementioned pair of marine-debris sunglasses, which cost $199 and are advertised as recyclable. The stylish shades are stamped with a unique QR code that when scanned reveals the origin story of the plastic litter that went into the sunglasses. Similarly, German-born designer Cyrill Gutsch, founder of Parley for the Oceans, collaborates with major sports and fashion brands such as adidas to design high-end apparel and items with components made from plastic collected at beach cleanups. "Purpose is the new luxury," Parley states on its website.

Modern artists also collect plastic and other kinds of marine debris, a now free and widely available creative medium. A few that immediately come to my mind include: Pam Longobardi, who creates enormous collages of arranged plastic artifacts, each carefully tagged with the name of the shore from which it was pulled; Cindy Pease Roe, who weaves intricately detailed, sometimes life-sized, marine animals made of plastic debris found on beaches near her studio on the East End of Long Island, New York; Swaantje Güntzel, who surrealistically portrays the consequences of the plastic crisis in a wide variety of media, including—in one work—glittery plastic microbeads extracted from beauty products, which she used to coat the body of a dead fish that likely had consumed microplastic in its lifetime.[13] These three artists have shown their work, made of others' trash, in prestigious international forums.

By using plastic litter as a resource, creators have brought conversations about the plastic crisis into the art world—inviting gallerists and viewers alike to consider their own relationships to plastic and prompting other artists to consider wastefulness when they create. Many art media—such as acrylic paints, foams, adhesives, fabrics, and the packaging they're sold in—are made of plastic.

In Denmark, I met another kind of artist, a storyteller, named Rasmus Holm. Holm is cofounder of a nonprofit called Skraldejagt: a Danish play on words translated in English to "trash-ure hunt." When we first met in 2018, Holm explained he had just begun organizing local trash cleanups across Copenhagen. He set out to gamify the process of picking up litter, encouraging people to search together in teams and to create stories about the provenance of their findings.

"Each of us lives in an exciting and unique local environment, but rarely do we get the opportunity to intimately get to know it," Rasmus explained. "We can learn a lot about ourselves as humans through deep exploration of our surroundings. Each bit of trash we find is connected to at least one person. Each bit of trash, like each person, has a story to tell."

Among the more remarkable finds Holm and his volunteers have made across Copenhagen include a flat-screen TV, presumably stolen; a suitcase, stuffed with shoes; piles of condoms, used; love notes, hand-written; wallets, emptied; and all manner of single-use plastic and paper items. Occasionally I've seen one of Rasmus's cleanup volunteers pocket an item, like a dirt-smeared compact mirror or a knockoff silver necklace. What is found—in Copenhagen, often wedged beneath shrubs, submerged in ponds and the city's central lakes, or half buried in flower beds—and what we do with it does indeed reveal much about human culture and how we relate to objects, plastic and otherwise, that come into our possession.

Getting hands-on with garbage can help change our relationship to plastic and other kinds of trash people create, encouraging us to be less wasteful and compelling us to continue picking up after ourselves. Soon I would learn how cleanups also play a role in changing our relationship to the Earth and to one another.

I met Rich Cramp in Nai Yang, one of Phuket's highly visited tropical beachside strips. It was Thailand's rainy season, and so the beach was nearly devoid of those people easily identified as tourists, by the looks of their wide straw-brimmed hats, dark sunglasses, dangling DSLRs, and deep sunburns. Cramp, who is a father and history teacher, was affiliated with Trash Hero, a Thailand-born, but now global, nonprofit focused on addressing the plastic crisis with beach cleanups, among other efforts. We settled into a pair of wobbly white plastic chairs bookending a round plastic table, on top of which someone had tossed a few laminated menus.

In our pre-meeting correspondence, Cramp had mentioned to me a specific obstacle he had faced in recruiting volunteers for his organization's cleanups. It was a challenge revealing of one human subculture's relationship with plastic litter, and an unjust consequence of the material's ready-made disposable status.

"It is a huge cultural taboo in Thailand to clean up other people's trash," he explained further, at our meeting. As Cramp and I spoke, I noticed a young man stepping around our table, his bare feet leaving shallow footprints in the sand. He proceeded to weave through the restaurant's collection of plastic furniture, sweeping with a long-handled straw broom, covering wayward plastic bottle caps, food wrappers, and utensils with sand.

In Thailand, it's the visitors who generate much of the country's trash. Nearly ten million people visit the island of Phuket annually, each one accompanied by the standard tourist's stream of single-use plastic go-tos, like mini containers of hand sanitizer and plastic water bottles.[14] Phuket also has a large expat community, many holding passports to the US—where the per capita plastic waste footprint is largest in the world. Rich expressed that over time, with enough cleanups, people in Thailand—visitors and locals alike—will hopefully stop considering litter as "someone else's problem."

We know that when plastic is not cleaned up it will stay on the ground, or be carried by animals, wind, or water, often to the ocean. I spoke with a few Trash Hero volunteers who are Thai, and they confirmed the existence of a trash taboo. Most agree it seems to have emerged from Buddhist beliefs misconstrued to justify the poor as deserving of lower status. Those who are struck by challenges like financial poorness may be judged for seeming to have brought on such misfortune themselves—for piling up moral demerits during past lives, perhaps—though of course living in poverty is not a thing many people can easily control.

Compounding this dilemma is beachside businesses' widespread employment of beach cleaners, like the man I saw at Nai Yang, who make it seem like local litter is being dealt with—by others whose jobs involve sweeping plastic out of sight.

Those native Thai people I discussed this matter with—the cleanup volunteers; the hostess at my hostel, Somo; and beachgoers I met on my daily excursions to Phuket's beaches—expressed the same feelings we all do when faced with uncontained plastic waste, recognizing it isn't supposed to be there. When we find plastic litter, we each must answer an internal question: Will I pick it up?

In Thailand, as everywhere else, some people are willing to, others are not. The difference is, many of those who choose to do it in this region are doing so while defying what picking up others' trash means culturally.

This time of year, more likely to bring torrential monsoons than steady sunshine, Phuket was temporarily relieved of an additional input of much tourist trash. And yet, plastic proliferated Phuket's landscape: Plastic bags held all manner of food—sizzling just-grilled skewered squid, fresh pink dragon fruits, and capsaicin-spiked green papaya salad—sold in the bare bulb–illuminated stalls of Phuket Old Town's bustling night market. Plastic toys—the cheap, excessively colorful and flimsy kind that survive just a few days before a child destroys, loses, or tires of them—were stacked in open-air shops along Tambon Sakhu's main drag, Route 4031.

Convenience-store cashiers reflexively slipped every purchase, no matter how small, into a plastic bag, often with a plastic spoon or straw. Plastic was heaped into mountainous piles according to form and function—bottles, bags, car parts, appliances, buoys—behind Phuket's small houses and along the fences surrounding them, awaiting some undetermined fate.

Shiny plastic sachets once carrying single portions of spices, instant coffee, shampoo, and dish soap were scattered along Phuket's roadsides and winding dirt paths. Torn plastic bags tumbled into low-lying green shrubs and flew into tall green trees. When traversing this area of the island, one was frequently accompanied by scraggly hens, split-eared cats, and barking dogs. Humble stucco homes festooned with ornate red-and-gold tapestries were adorned with golden shrines holding fruit and bottled water, offerings laid before cross-legged Buddha statues. The homes' outdoor terraces were tight but tidy, their minute tiled expanses accommodating rusting washing machines, racks covered by drip-drying clothes, and a lineup of sandals belonging to each dwelling's occupants. At the end of the homes' dirt driveways, mini landslides of rubbish—mostly plastic—cascaded from overfilled yellow bins.

Amid the plastic, rain drops, and palm fronds, human shapes—an eye, a hand, a mouth—would emerge. People smiled and waved along the roadsides and from the open windows of their homes. "Sawasdee," they called. Hello.

Thailand's plastic problem—by the looks of endless trash heaps; blatant littering; not-so-secretive illegal dumping; and prolific consumption of single-use plastic bags, straws, bottles, and other items I encountered in Phuket—was astounding. In fact, by 2020, scientists would declare Thailand the world's fourth most prolific coastal contributor of so-called mismanaged plastic waste to the natural environment, after the US, India, and the world's worst offender, Indonesia. The US stands out in the list's top five as the only high-income country, and one of the world's biggest exporters of plastic waste to countries lacking capacity to recycle it—including Thailand.[15]

The road from Nai Yang wound through quiet rural villages and sprawling suburbs to the province of Rawai on the southern tip of Phuket. From the open window of a stuffy cab, leather seats slippery with sweat and humidity, I watched messy piles of garbage, unlawfully dumped, festering along forested roadsides. Thinned trees indicated clearings, from which streams of choking black smoke bellowed up into the blue sky. The putrid stench of singed garbage frequently prickled my nose. There was far more garbage inland, here, than around Phuket's paradisiacal beachside resorts—cleaned by those hired to haul it out of most tourists' lines of sight. As the cab approached the coast and Rawai's beaches, hotels, and restaurants, trash sightings grew less frequent—though not infrequent.

On Nai Harn Beach, around fifty people—locals and foreigners; men and women; seniors, adults, teens, and children; intense athletes and self-described couch potatoes—sprinted together, herd-like, across the sand. British expat Krix Luther, a broad-shouldered fitness instructor, stood by the shore break shouting out instructions to the crowd as gentle waves grazed his bare toes. The mass of exercisers churned like a swarm of bees—moving, always moving: running and crawling and jumping across the beach's soft sand and splashing through its turquoise-blue water. When most people were red faced and out of breath, Luther cued a cooldown and carried boxes of blue latex gloves and black polyethylene garbage bags to the group. (When I asked why they were cleaning up with more plastic, Luther said, "It's what we can afford right now.")

This was no ordinary beach boot camp, it was Luther's Clean the Beach Boot Camp, another nonprofit cleanup group created to address Thailand's growing plastic problem.

"One day we were training on Nai Harn and saw the worst amount of trash I have ever seen," said Luther, who moved to Thailand in 2008 to fight Muay Thai. "In fact, I was afraid someone was going to get hurt.

So I stopped the workout forty-five minutes in and asked everyone to clean for twenty minutes. They all had huge smiles on their faces afterwards, from the adrenaline of the workout and satisfaction of cleaning. Now we're up to exercising for an hour, and cleaning for an hour. It's a solid workout."

After some stretches, the cooldown was complete, and the exercisers rose, sand clinging to their sweat-soaked skin and clothing. One man grabbed hold of a thick tangle of fishing ropes, tattoos rippling across his muscled shoulders as he hauled the heavy knot across the sand. A mother and son waded through a lush patch of green scrub, plucking out plastic bags and wrappers with gloved hands. A young woman scoured the reedy wrack line for plastic bottles and caps (rarely attached to one another), plastic utensils, and single-use sachets. They carried the trash, stuffed into the black garbage bags, to the beach parking lot for a weigh-in. The total heft of the day's trash haul was about 660 pounds.

As the volunteers piled their bags next to the single small rubbish bin standing in the beach parking lot, Luther and a few men installed two additional garbage receptacles, made of bamboo, on the beach. "I hope the bins send a visual signal to people to properly dispose of their trash rather than leaving it on the sand," Luther said.

"Where does the trash go after someone throws it into a trash can?" I asked, watching Luther smash the bamboo poles down into the sand with a hefty rock.

Luther paused. "Well, that's another problem. We don't really know. We are told the municipality of Rawai picks it up, and they bring it to Phuket's incineration plant."

"Or they could just be tipping it back into the ocean," one of the men helping Luther shrugged. "Sure seems like it. We found a refrigerator washed up here last week. I mean, how the hell did *that* get *here*?"

Though the morning's workout and cleanup had come to an end, several participants lingered, continuing to comb the beach for trash.

This included a woman named Nangy Phanchana, a Thai freelance tour guide. She noted that plastic's rise in Thailand coincided with the country's recent international tourism boom. "Even just fifteen, twenty years ago, much of the trash we'd find on local beaches was banana leaves because we knew they'd disappear in time," said Phanchana. "That was before tourists started pouring into Thailand in the late 1980s."

Emanuele Mario Montalde, a young Thai man who had just graduated high school, was hunched nearby, listening, as he sieved microplastic from the sand with his fingers. He agreed with Phanchana: "Thailand began using plastic much later in the game than many other countries—in my grandparents' and parents' generations they used materials like banana leaves, glass, metal, and paper to hold food and make things."

As a budding environmental conservationist, Montalde has tried to investigate where the trash goes. "It seems the state contracts some cleanup efforts and resorts will pay people to pick up trash," he said. "But there's not a great infrastructure in place in Thailand to deal with plastic today. And from what I have learned, there's not really a great infrastructure anywhere."

CHAPTER 11

# Closing the Loop

As we've seen, removing plastic from beaches and waterways is a monumental task that doesn't come close to solving the crisis on its own. But the wave of public awareness that cleanups have inspired is now helping push forth critical efforts to reduce the amount of plastic we produce and transform it from a throwaway material into a valuable commodity that is always reused.

As awareness of the plastic crisis rises, so does demand for more eco-friendly single-use products. One commonly sold replacement to plastic is the more ecologically benign sounding "bioplastic." However, much "bioplastic," made entirely or in part from plants or bacteria instead of petroleum, is no better for the environment than its synthetic counterpart: The plants most commonly used to make bioplastic include corn and sugarcane, two crops closely linked to the proliferation of deforestation, fertilizer and pesticide use, wasted water, and soil degradation.[1]

What's more, once used, bioplastic products must still be collected and processed for landfilling, incineration, recycling, or composting. Unlike conventional petroleum-based plastic, which fragments over time but forever remains plastic, some bioplastics can indeed degrade into simpler chemical components. But this doesn't happen particularly rapidly and, in the meantime, bioplastic products can pose a hazard to wild animals, which may ingest or become entangled in them—just like conventional plastic products. Recyclers lament that bioplastic items—which look, feel, and act a lot like those made from conventional plastic and so are commonly discarded as such—are increasingly contaminating their stocks of recycled petrochemical-based plastic. And when landfilled, dumped, or incinerated, bioplastic's breakdown also releases greenhouse gases into the atmosphere, ultimately contributing to climate change.

"If a container made of PHA, a type of bioplastic made of bacteria, found its way on the side of the road, there it would degrade into carbon dioxide, methane, and water within two years," Dr. Joseph Greene, a materials scientist and professor at California State University, Chico, has explained to me. "But with plant-based bioplastic, like PLA, it could take up to ten years to decompose [in nature]."[2]

Certain types of bioplastic, however, can be composted under strictly controlled environmental conditions—ideal oxygenation, heat, and humidity—in commercial composting facilities, a trait conventional petrochemical-based plastic does not possess. A growing number of American cities—such as Boulder, Colorado; Portland, Oregon; and San Francisco, California—have commercial composting facilities and offer residents curbside pickup of compostable items. San Francisco is credited with developing the first large-scale urban composting program in the US. The city's waste management company, Recology, reports having diverted more than two million tons of food scraps, food-soiled paper, wood, yard trimmings, bioplastic products, and other biodegradable items from landfills to its composting facilities since the program started in 1996, saving the atmosphere a significant amount of climate-warming greenhouse gases.[3]

With landfilling, organic matter is piled up to rot without any added oxygen. Without oxygen, decomposing organic matter releases plenty of carbon dioxide as well as methane, a greenhouse gas twenty-six times more potent than carbon dioxide.[4] Compost, on the other hand, is turned while it's breaking down, a method that introduces oxygen, encouraging decomposition of organic materials and releasing a significantly smaller amount of carbon dioxide as a byproduct. What's more, composting doesn't require ever-growing space like landfilling does. And it's actually useful, seeing that compost can fertilize crops and gardens. Compostable bioplastics are just one of many plastic alternatives.

"It seems like the most natural material placements hold the most promise," said Theanne Schiros, assistant professor of math and science at Fashion Institute of Technology (FIT) in New York City. At FIT, Schiros and a group of her students have turned to nature in an attempt to put wastefulness and pollution in the apparel industry out of fashion by growing the components of biodegradable textiles from live organisms. The result: eco-friendly clothing materials, some grown into near-complete items without the need for factory assembly.

"Tomorrow's clothing could be made from living bacteria, algae, yeast, animal cells, or fungi," Schiros told me over video chat, above the din of her students, whom I could see scurrying around her lab. That day, they were working to develop yarn made from algae and dyes with nonchemical pigments such as crushed insect shells.

At one point, she held up a sheer white tank top to her camera. It looked lightweight and comfortable. "This is made of algae," she explained. "When thrown away, this will break down into harmless substances that could be reused by nature."

This, instead of breaking up into tiny plastic microfibers and fragments that easily enter the natural environment, as conventional synthetic apparel does. These textiles made from natural materials and living organisms are so far mainly constrained to the laboratory, science competitions, and high-fashion runways. But Schiros believes it is just a matter of time before such innovations are introduced, in some form, to consumer markets. The first step is to create natural clothing durable enough to stand up to all the usual wear-and-tear mainstream apparel

is subject to. Schiros sees potential in using Indigenous preserving techniques—such as tanning with smoke, instead of chemicals—which she
said can lend materials like "bio-leather" (a vegan leather grown from
a liquid bacteria culture, fungi, and compostable waste) strength and
water resistance. In 2019 she cofounded Werewool, a textile biomate-
rials company, to explore such opportunities.

Other proponents of nature-based clothing point out that, in order
to succeed, such apparel needs to be cost-competitive with conventional
clothing. For example, sustainable textile innovator Laura Luchtman,
founder of Dutch design brand Kukka, sells bacterial-dyed silk scarves
for about $130 each, whereas a similar silk scarf dyed conventionally
with synthetic dyes can be purchased for as little as $10 each, though the
cost to the environment is much greater. Natsai Audrey Chieza, another
sustainable material pioneer, is also developing bacterial dyes as founder of
London-based biodesign lab and creative research agency Faber Futures.
"Similar to the debate around renewable energy, cost-competitiveness
will not only rely on solid science and a technology that works—it will
need to be enabled through government subsidies and a mental switch
towards investing in R&D," Chieza explained.

Beyond fashion, innovators are increasingly tapping bacteria, fungi,
algae, and other living organisms for their plastic-replacing potential.
Some plastic alternatives are astoundingly simple: In 2019, Rimping
Supermarket in Chiangmai, in northern mainland Thailand, stopped
wrapping its produce in non-recyclable plastic film and started using
banana leaves instead. Like plastic, banana leaves are inexpensive. But
they're also biodegradable, abundant in Thailand, and—as I learned
during my visit to Phuket—were in fact long used as throwaway food
packaging in Southeast Asia . . . until plastic came along.[5]

In 2007, Eben Bayer and Gavin McIntyre, who had developed and
patented a mushroom-based insulation called Greensulate, went on to
cofound a New York–based company called Ecovative. Their team developed an affordable, lightweight, thermally insulating, and water-resistant

packaging material made from hemp and mycelium—the vegetative spiderweb-like substance that produces mushrooms. It easily replaces Styrofoam in form and function, yet breaks down benignly in less than thirty days when composted. From its flagship Mushroom® packaging, Ecovative has expanded its mycelium-based offerings, selling completely biodegradable makeup sponges, leather, and footwear, and even plant-based "meatless meats" created to combat animal cruelty, deforestation, factory farm pollution, and climate change.

And then there are metal (usually steel or aluminum), glass, and paper—plastic's biggest competitors as favorite mass-produced materials of the Industrial Age. While these materials have a better track record than plastic when it comes to being recycled, they come with their own ecological trade-offs, especially when produced on a scale fit for mass consumption: toxic pollution and related environmental justice issues, deforestation and loss of wildlife species, burning fossil fuels for energy to power production and recycling.[6] And yet, one single stainless steel water bottle, if used throughout the day, replaces thousands of plastic water bottles potentially discarded in one year.

Today, a growing mixture of businesses, nonprofits, and governments are forging ahead with plans of circularity in relation to plastic and conserving nature. Loop, a subsidiary of recycling company TerraCycle, is partnering with major brands to establish a marketplace where all kinds of goods from ice cream to liquid hand soap are sold in containers that are returned and refilled (think, the modern milkman). Replenysh, a software company, is building digital tools that companies can use to better recover and reuse the waste their products produce.[7] by Humankind sells natural personal care products, like toothpaste tablets and deodorant, as refills and in their refillable containers, many of which are made of recycled plastic.

Many business owners are quick to admit that their recent adaptation of a more circular ethic is driven by popular demand. "Great innovations are happening as customers raise the bar on expectations for sustainability and expect more from brands they buy from," said *by* Humankind cofounder and CEO Brian Bushell. "These new ways will become more mainstream over time, and hopefully not too much more time—hopefully time enough to save the planet."

Recent efforts to circularize are part of a more natural, equitable, and intelligent way forward. Everyone—governments, businesses, entire industries, nonprofits, and individuals—needs to be involved in shaping a new, more circular existence. And clearly, the only morally sound and survivable strategy is to balance our use of the planet with its protection. As explained by economist Kate Raworth, humanity, fixated for so long on using the empty measure of GDP to track human progress, needs a new metric—and that is "meeting the human rights of every person within the means of our life-giving planet," Raworth writes in her book *Doughnut Economics: Seven Ways to Think Like a 21ˢᵗ Century Economist* about a radical, but intuitive, plan to circularize the economy.

It's not just Raworth calling for a rounding of our highly exploitive economic system: The World Economic Forum has supported initiatives that push forward the circular conversation, and new sustainable economic developments, supported by dozens of nonprofits, like the Ellen MacArthur Foundation—founded in 2010 by English sailor Dame Ellen MacArthur—and even some corporations, like health-technology giant Royal Philips, headed since 2011 by CEO François Adrianus "Frans" van Houten.

While mainstream economics uses a circular flow model to show how money moves between businesses, governments, households, and financial institutions, Raworth's redrawn doughnut-shaped economic model adds nature, energy, and unpaid work into the classic oversimplified, inadequate economic system. The "doughnut"—an inner ring representing the things humans need to survive, like food, clean water, housing, education, health care, and democracy; an outer ring marking Earth's natural limits on these human needs; and a center representing a position of deprivation—shows us where we need to strive to live: in the sustainably habitable space between the two rings, where a circular economic model could allow us to thrive.

As Raworth explains, "Far from being a closed, circular loop, the economy is an open system with constant inflows and outflows of matter and energy. The economy depends upon Earth as a source—extracting finite resources such as oil, clay, cobalt and copper, and harvesting renewable ones such as timber, crops, fish and fresh water. The economy likewise depends upon Earth as a sink for its wastes—such as greenhouse gas emissions, fertiliser run-off and throwaway plastic. Earth itself, however, is a closed system because almost no matter leaves or arrives on this planet: energy from the sun may flow through it, but materials can only cycle within it."[8]

The present call for a circular shift is not the first in recent human history. During the 1980s, Daniel Knapp, a former sociology professor fascinated by waste picking, developed an idea he called Total Recycling. Under Knapp's system, twelve commonly trashed materials, including plastic, are treated as valuable resources that would be sold back to businesses; nothing is simply thrown away or dropped in a bin.

In 1995, Daniel took his recycling ideas from California to Australia, where he collaborated with local governments and nonprofits to establish recycling facilities and systems designed to work toward a goal of "No Waste." Daniel worked with a local group of female waste pickers to help sort and sell recyclable materials in the waste stream, reducing the overall strain on Australia's landfills and maximizing recovery of useful resources. Later, he'd return to California and eventually join his wife, Mary Lou Van Deventer, in opening a resource-recovery business in Berkeley, called Urban Ore, which still sells all manner of discarded items and materials to the public out of an enormous warehouse. Throughout the 1990s and early 2000s, the internet helped spread the word about Knapp's No Waste ideals, which were adopted in environmental circles around the world, and gave rise to the present "Zero Waste" movement.[9]

Wasting nothing involves a conscious rethink of how humans value materials. Whether or not we are aware of it, most modern people possess at least a few Zero Waste habits, such as buying or donating used items at flea markets, hosting yard sales, swapping clothing, and composting kitchen scraps. Public polls reveal many people have a preference for buying products made from recycled materials.[10]

The Zero Waste movement's biggest hurdle is getting people to step out of their convenient comfort zones and give living less wastefully a chance. And as habits shift and Zero Waste becomes more mainstream, the movement has sometimes struggled to stay focused on its primary goal without falling prey to commoditization. In the past, Zero Waste proponents tended to be dumpster-diving folks carrying a certain '70s counterculture aesthetic. At present, some of the most visible faces of the Zero Waste movement are predominantly young, educated, and privileged—characteristics not problematic but indicative of a trend and possible challenge.

Across the internet, self-described "influencers" fill blogs and social media feeds with plugs for often-pricy reusable products like silicon ear swabs and foldable metal straws (for which many receive sponsorships). For one, many people cannot afford these products. That gives rise to the possibility Zero Waste will continue to carry the same kind of elite environmentalism we've seen with electric vehicles and organic food—things pricier than their decidedly less ecologically friendly alternatives. At the same time, the objective of Zero Waste is to use less stuff, period—so buying all manner of fancy reusable items is a bit contradictory.

To be successful, the Zero Waste lifestyle must be accessible to all people. But even as some people change their personal habits, many more people will continue to use plastic. And, as we know, corporations plan to only produce greater amounts of it into the future—despite our need to make vastly less plastic, not more.

If left to their own devices, petrochemical and plastic corporations will keep making plastic until Earth's fossil-fuel stores run dry. That is unless some other entity requires them to stop. Historically, many governments—including those of the US and EU nations—have weakened the petrochemical and other chemical industries' ability to mass produce other harmful substances, like the toxic insecticide DDT, by enforcing regulations on chemical manufacturing, use, and sales.

In the 1960s and '70s, public opposition to DDT—greatly amplified by the voice of ecologist Rachel Carson in her 1962 book *Silent Spring*—grew so overwhelming that after less than three decades of use, bans on DDT manufacture and use began cropping up around the world. On May 22, 2001, 92 countries adopted a global treaty called the Stockholm Convention on Persistent Organic Pollutants to severely restrict use of DDT—allowing limited use only when needed to kill mosquitoes spreading malaria—and completely banned ten other chemicals around which health concerns had arisen. Today, 152 nations (including most members of the EU) are signed onto the Stockholm Convention, which has adopted amendments restricting or prohibiting the manufacture and use—and sometimes requiring cleanup—of more than two dozen highly toxic chemicals once mass-produced and sold by industries.[11]

One of the newest classes of chemicals to make the Stockholm Convention's list are perfluorooctanoic acids (PFOAs), which are considered PFAS—those petrochemical-based additives commonly used in plastic, nonstick "Teflon" cookware, and firefighting foam. PFOAs, we now know, cause a variety of severe health problems in people and other animals, most notably related to reproduction and hormone function—and are now ubiquitous in our environment as a result of their historic and continued production and use.[12] PFOAs, DDT, and many other chemicals covered by the Stockholm Convention, classified as "persistent organic pollutants," do not degrade quickly or benignly in nature. And so the legacies of now-restricted chemicals perpetuate; indeed, they circulate air, water, soil, and even our bloodstreams. Still, the human health and ecological outcomes of the convention have been generally positive, particularly in the case of DDT—the widespread ban of which has been linked to a significant rebound in the populations of several wild bird species, like osprey and bald eagles, that were nearly wiped out by the chemical.

While it's useful to regulate dangerous plastic additives and other toxic chemicals produced by big industries, what about plastic itself, any piece of which could serve up a cocktail of thousands of chemicals—including those known to harm us? In 2019, UN delegates met in Nairobi, Kenya, to discuss a proposed phase-out of all single-use plastic items by 2025. But the talks ended inconclusively, with some member states—particularly those with significant investments in fossil fuels and plastic production, like the US—voicing loud opposition. In the end, UN member states agreed on a vaguely worded, nonbinding commitment to "significantly reduce" use of disposable plastic items by 2030.[13]

Shortly after the UN summit, David Azoulay of the Center for International Environmental Law, a nonprofit focused on human rights and environmental justice, reflected, in an interview with Reuters, "The vast majority of countries came together to develop a vision for the future of global plastic governance. Seeing the US, guided by the interests of the fracking and petrochemical industry, leading efforts to sabotage that vision is disheartening."[14] At another UN summit the following year, more than two-thirds of UN states expressed willingness to participate in the creation of a global agreement to curb plastic production. Again, delegates from the US—and several other countries with prolific plastic use, like the UK—declined to support such a pact.[15]

While world leaders have not yet passed a binding global treaty restricting or banning use of plastic, the possibility seems increasingly likely moving forward. By now, many people have heard of plastic bag taxes and bans, and other rules limiting local or national availability of single-use plastic items. Since Denmark became the first country to pass legislation taxing plastic bags to disincentivize their use in 1993, legislation curbing single-use plastic has sprouted up around the world. Today, more than one thousand different laws restricting or banning single-use plastic items have been passed in municipalities, states, and nations.[16]

Currently, the continent of Africa is leading the world with the highest number of national rules for single-use plastic bags: Thirty-four African countries now have strict laws against using, making, and importing plastic bags and, in the case of Rwanda, all single-use plastic items. In Tanzania, visitors traveling to the country through high-traffic entry points like airports are now required to "surrender" their plastic bags at designated drop-off points.[17] Run afoul of the rules and risk heavy fines and/or jail time. While actual enforcement of these laws varies, loopholes allowing certain applications of single-use plastic exist, and anecdotal reports say money can help sway officials to look away from violations. But for the most part, the UN has found plastic bag bans effective at curbing local plastic pollution and the common (and hazardous) practice of burning plastic waste.[18]

The US, by comparison, has successfully passed just one strong piece of national plastic legislation: the Microbead-Free Waters Act, which was signed into law in 2015 by President Barack Obama and banned the manufacture and sale of products containing plastic microbeads in phases after 2017. As a result, while many companies complained, major brands like L'Oréal and Unilever ultimately agreed to remove the pesky plastic beads from their products and replace them with more ecologically benign exfoliating alternatives, like ground-up nut husks and plant waxes.[19] While the microbead ban has largely been praised as a success, a worrisome trend is now emerging in the US in response to the rise of proposed single-use plastic legislation. Across the country, particularly in those states with majority Republican representation, the phenomenon of statewide "preemptive" plastic laws that make it impossible to pass restrictions on single-use plastic—essentially, bans on plastic bans—are on the rise. By 2020, more than a third of US states had passed or were trying to pass preemptive plastic legislation strongly lobbied for by petrochemical and plastic industry trade groups representing corporations making plastic and its ingredients.[20] And so, in the US, single-use plastic legislation has tended to most successfully pass on municipal levels, as it bypasses some of the bureaucracy of state and federal politics.

Yet, some states have managed to pass fairly strong single-use plastic policies. The first to do so was Hawai'i, which passed a statewide law banning plastic bags in 2012.[21] While a concerned public largely supported the legislation, Hawai'i's road to reduction—like so many other states' journeys—has been long and contentious, and fiercely opposed by industry.

In 2017, Honolulu Mayor Kirk Caldwell signed Bill 59, legislation designed to close a major loophole woven into O'ahu's plastic bag ban since it was first implemented in 2012. Despite passage of the statewide bag ban, O'ahu, Hawai'i's most-visited and most-populated island, still permitted retailers to continue giving away plastic bags, so long as they were slightly thicker than typical single-use grocery bags because they were considered "reusable." The ban was so patchy that it still permitted takeaway establishments and farmers' markets to distribute the lightweight single-use plastic bags the state was trying to eliminate.

Hawaiian residents and scientists observed that people were treating the thicker "reusable" plastic bags—made of slightly heavier plastic more resistant to tearing—the same as the lighter single-use plastic bags they replaced, as evident through the alarming number of plastic bags still found strewn across the island after the legislation was passed. Under the revamped rules, retailers would be required to stop distributing bags—of any kind—for free, even the thicker reusable plastic bags exempt by the former loophole, by 2020. In a bid to push customers to bring their own reusable bags, shops were required to charge fifteen cents to buy a paper or thick plastic bag.

Yet exemptions meant to appease makers and sellers of plastic remained, and retailers selling one or more items detailed on a long list—including plants, fish, frozen foods, prepared foods and beverages, baked goods, and medications—were still permitted to give away plastic bags for free to hold those items that were considered too messy or fragile to change hands from seller to consumer without a plastic wrapper.[22]

Clearly, although bans on single-use plastic items may seem simple, that doesn't mean they're easy to pass everywhere. I observed an intense battle over another commonly littered plastic item on Oʻahu while helping prep *Christianshavn* for her journey from Honolulu to Nuku Hiva in fall 2017. Since a plastic bag ban was already in place, the next plastic target in sight was expanded polystyrene (EPS) foam, better known as Styrofoam, which is ubiquitously littered on Oʻahu's streets and beaches.

"We have sailed our way from Europe to here, taking samples of the ocean all the way and what we see as we get nearer Hawaiʻi is more and more Styrofoam," Torsten testified before Oʻahu's city council members, who sat deliberating over a potential ban on Styrofoam containers across the island. "It's a good message to send to the rest of the world that hereby with this bill, we will stop this pollution; that this is the way to go forward for coming generations."

He finished and, nodding to the council in conclusion, rose from his seat and squeezed across the packed meeting room to join me in the back where I was standing by the door. He'd been one of the last in the room to testify that day in October 2017 at Honolulu Hale, city hall, and, like the dozens of other people who spoke to the council, was allotted just a few minutes to make his case.

On one side of the meeting room sat plainclothes scientists, flip-flop-clad surfers, and stay-at-home parents. On the other, restaurant owners, and food industry reps, and employees gathered. The hearing attendees, either wearing faces of determined support or exasperated opposition, overflowed from a few neat rows of folding chairs up to the doors of the small room waiting for their turn to be heard. Forced together in the same small space, the pressure was rising in an already shaken bottle, one that threatened to burst.

While many statements blurred into one another, it became clear that many more people supported the legislation, Bill 71, than opposed it—and those who did oppose it almost exclusively owned, worked for, or represented companies that had something to lose—namely, money— should a ban on Styrofoam, or EPS, containers come to pass. Opponents largely suggested plastic wasn't a problem—*littered* plastic was, and should be addressed through better litter-prevention schemes, not outlawing EPS foam altogether. This, though many supporters of Bill 71 pointed out there wouldn't be so much litter if less plastic was being handed out to people in the first place.

Only one person, an O'ahu resident named Kirk Markle, expressed his opposition to Bill 71 for a completely novel reason: his personal aversion to eating a plate lunch out of any container other than plastic. He testified in favor of leaving diners the choice to eat out of plastic, as compostable containers often became soggy in his experience. (As it happens, plate lunches weren't always served in plastic. The immigrant laborers on Hawai'i's sugar and pineapple plantations consumed the first plate lunches—their previous night's leftovers, beefed up with heaps of white sticky rice—from wooden bento boxes. Later, lunch wagons sold these meals on disposable compartmentalized paper plates, finally lending it the nickname "plate lunch." Next the paper plates were traded for the foam clamshell containers—and here we are today.)

A man named Ari Patz, who had shaken his head during Markle's testimony, later took the stand and introduced himself as a representative of World Centric, a company manufacturing compostable single-use food and beverage containers. "I will challenge anyone's beef stew or saimin or ramen or anything in any of our containers. I pretty much guarantee you they're going to hold up," he said.

Patz held up one of his company's compostable clamshell containers, which was brown and earthy looking. He explained it was made of stripped harvested wheat stalks—something that's usually an agricultural waste material. So instead of lasting indefinitely in the environment like EPS foam containers, the compostable clamshells break down into simple natural substances like water and carbon dioxide within 180 days in a commercial composting facility. However, Patz acknowledged, such a facility is yet to be built on Oʻahu.

"I believe that the ideal end of life for this particular product would be a continuation of its life, which would be a full cycle," Patz said.

He suggested Oʻahu ramp up existing small-scale composting operations and establish a "one-bin solution," where homes and public spaces could have designated baskets for discarding used compostable containers and utensils as well as food waste. The biodegradable contents of these baskets would be sent to a composting facility to be turned into rich soil, something useful—even desirable—by golf courses and farms, instead of being thrown in a landfill where they release climate-warming gases as they break down over time. Ari added that Oʻahu's waste audits have revealed at least 60 to 70 percent of the island's waste as potentially compostable. The anticipated influx of compostable containers into the waste stream after passage of an EPS foam container ban like Bill 71 would only make such a venture more lucrative as more compostable material is expected to be added to the island's waste mix.

There is currently no infrastructure for recycling EPS foam across the state of Hawaiʻi and EPS recycling facilities and programs exist patchily across the mainland United States. (Though it's worth noting that EPS recycling rates are dismal even when facilities are available because it's hard to cleanse used foam containers of food residues, and they often crumble into bits that gunk up recycling machinery for other types of plastic.) So, O'ahu burns its EPS foam and other kinds of waste at H-Power, the island's waste-to-energy facility, or sends it to a landfill, leaving people with only one option for responsible disposal: tossing foam into the trash.

When burned, EPS foam emits more climate-warming greenhouse gases ton-per-ton than most other materials and even any other kind of plastic.[23] If that wasn't bad enough, EPS foam also releases toxins, including styrene—a chemical suspected to cause cancer and reproductive issues, and known to cause skin and eye irritation, digestive ailments, and neurological problems in people—at all points in its life cycle, from when it is first produced, used to hold food or beverages, and then burned, buried, or littered in nature.[24] Wild animals—particularly those living in and around the ocean—commonly swallow EPS foam after mistaking it for food, with often-lethal results.

These are the consequences of our perceived convenience of, and consequent reliance on, EPS foam—a material used for packaging, food and beverage containers, building materials, and even children's toys. It took several decades for plate lunch to become iconically linked to its foam container. Perhaps it will only be a matter of time before foam defenders can get used to eating plate lunch out of a more sustainable vessel—perhaps the classic reusable, refillable container, or a new single-use bioplastic clamshell.

World Centric's products are certified biodegradable by several third parties, including the Biodegradable Products Institute in New York, and it claims all its products can be composted commercially in two to four months.[25] Proponents of biodegradable alternatives to plastic say compostable containers—even in a landfill, even in an incinerator, even in the ocean, even on the side of the road—appear less harmful to plants and animals, including people, and the planet than the plastic containers they are meant to replace. Yet scientists are challenging these new materials and caution against so-called greenwashing: marketing meant to convince people that a product or service is sustainable.

"Bio-based and biodegradable plastic are not any safer than other plastics," said Lisa Zimmermann, a PhD student at Goethe University in Frankfurt and collaborator with plastic research group PlastX. In 2020, the results of an experiment she led—testing the toxicity of forty-three single-use bioplastic items, such as disposable cutlery, beverage bottles, and wine corks—were published. Zimmermann and her team found more than one thousand different chemicals—including some toxic additives commonly used in petrochemical-based plastic—in 80 percent of the products. A few contained up to twenty thousand different chemicals. And like producers of conventional plastic, those companies making bioplastic tend to lack transparency about their proprietary ingredients.[26] What's more, bioplastic products are often indistinguishable from the real thing, leading people to "wish-cycle," placing nonrecyclable items in recycling bins—a practice grounded in good intentions but also one that can botch a whole batch of recycled plastic.

As demand for biodegradable single-use products continues to grow with the passage of new regulations, it will be critical for companies to make bioplastic and other plastic alternatives in ways that minimize environmental impacts, and without use of toxic chemicals. While more research must be done to outline the full range of ecological consequences linked to manufacturing biodegradable single-use products, present concerns make a strong case against throwaway items of all kinds.

Indeed, much legislation, particularly bag laws, disincentivizes use of not only single-use plastic but also single-use items made from other materials. New Jersey, which in 2020 passed the most comprehensive single-use plastic law in the US to date, tied a phase-out of disposable paper bags into its sweeping ban on single-use plastic bags.[27] Experts say efforts like this, which subtly shift the single-use mindset by requiring people to bring reusable bags or carry items without a bag, is crucial when formulating effective legislation.

This shift is already under way. At the meeting, Rafael Bergstrom, who then worked as Oʻahu coordinator of the Surfrider Foundation (and now directs the Hawaiʻi-based nonprofit Sustainable Coastlines), explained that their organization had already persuaded more than 140 local restaurants to voluntarily eliminate single-use plastic items, like EPS foam food containers and plastic bags, bottles, and straws. Across the US, other local Surfrider Foundation chapters were encouraging eateries to abandon single-use plastic products in favor of reusables and plastic alternatives—not necessarily just bioplastic, but natural materials like bamboo and banana leaves, too—as part of the organization's Ocean Friendly Restaurants program.

Similarly, UK nonprofit Surfers Against Sewage runs a program called Plastic Free Schools that supplies students, staff, and administrators with support to eliminate single-use plastic in the learning environment. "We see efforts to deal with plastic on every scale: seven-year-old kids who are trying to get their schools to stop distributing plastic straws or forks; people are getting their houses of worship and workplaces to cut their reliance on single-use plastic; all around the world we see cities, states, and some countries taking action; and even companies are taking steps to address consumer concerns," reflected John Hocevar, director of Greenpeace's Oceans Campaign. "The issue is: While most people agree there's a problem, we don't all agree on solutions."

While supporters of Bill 71 spent much time at the hearing outlining the health and ecological risks of using plastic, those working in Oʻahu's EPS foam industry seemed skeptical of their concerns. "I have worked at Hawaiʻi's Finest Products for fifteen years with no health problems," said Rey Ramos. "I don't usually get involved with politics. However,

Bill 71 will impact my job and family, so I am here." Michael Nakato, another one of the company's employees who showed up at the meeting, said, "I have not had any health problems working where I do, and the FDA has approved Styrofoam for use, so it must be safe."

After Nakato spoke, Councilmember Kymberly Marcos Pine, who proposed Bill 71, pointed out that the FDA has also approved cigarettes for use, but that they can cause severe, irreversible health effects. According to scientific evidence, so can plastic, she reasoned. A tax, ban, or other deterrent would work to quash demand for plastic, as it does for cigarettes. The more expensive or challenging it is to obtain a dangerous product, the less likely people are to get their hands on it and expose themselves and others to danger. And in the long run, reducing plastic litter could have meaningful cost savings for residents.

"Much of the testimony we have heard today [reflects] the cost burden from the business community and I want to point out that the public has a heavy cost burden," said Nicole Chatterson, who works at the office of sustainability at the University of Hawai'i and directs the organization Zero Waste O'ahu. "We're spending millions of dollars as taxpayers cleaning this stuff up. So that has to be considered when we're considering the price point of Styrofoam versus other types of containers."

Yet the oppositional voices of industry, though outnumbered, would make a clear impact on the collective psyche of the six-member council. After listening to forty-six testimonials, Committee Chair Carol Fukunaga expressed concern about Bill 71's potentially negative effects on local food establishments and EPS foam manufacturers. She said she preferred if strategies to reduce plastic waste and litter were motivated with incentives rather than mandates. She recommended, and other committee members agreed, deferring action on Bill 71 to make time for further discussions with all stakeholders, especially those who would potentially feel the economic aftershocks of the bill's passage—namely the owners of food establishments and employees at the local EPS foam manufacturing plant.

An exhausted-looking roomful of people emptied out of Honolulu Hale's fluorescent-lit halls into the bright light of the day. For the time being, EPS foam containers would continue to be used, burned, buried, and littered on Oʻahu—and blown into the sea.

Months later, Bergstrom pinged me with an update on Hawaiʻi's proposed foam legislation: The bill died before being heard in its final committee. Oʻahu's Surfrider Foundation and its allies would have to formulate another attempt to convince their local lawmakers it was necessary to take serious action on EPS foam, and other single-use plastic.

"Corrupt fucking world we live in," Bergstrom fumed.

Two years later, Nicole Chatterson, Rafael Bergstrom, and others who'd shown up at Bill 71's hearings were again speaking out against plastic in Honolulu Hale. This time, they expressed support for Bill 40, proposed legislation that would prohibit shops and food vendors from distributing most single-use plastic items—including the foam containers and cups targeted by the failed Bill 71, in addition to plastic bags, plastic straws, cutlery, and other single-use plastic foodware.

"This is going to make our businesses stronger for a future that is actually inevitable . . . and at the same time protect the places that we love and live in," Bergstrom said at a hearing in support of Bill 40. He told me he worried the bill might not pass, as the usual plastic proponents again swarmed city meetings to oppose further restrictions on Oʻahu's plastic use. Yet this time, lawmakers passed the legislation on the table—an even greater victory than Bill 71 could have ever been. Now the next steps can begin: creating the new processes, facilities, jobs, and mindsets that make up a truly circular waste system.

Recycling plastic has never been simple, but experts agree it's something we're going to have to get much better at doing if we want plastic to fit into a livable future.

"Plastic is a tricky material—it's more complex than, say, glass, which can be recycled over and over again into the same substance," Dr. Chelsea Rochman, a plastic expert at the University of Toronto, told me. "Most plastics lose some of their physical integrity while being recycled, creating a less-desirable, less-valuable end product. Plastics collected for recycling are also likely to be contaminated by debris and residues, as well as additives that lend color and other qualities. Recycling methods and systems need an overhaul if we're to seriously address the world's growing use of plastic—and its consequences."

A transformation of recycling is now under way. In recent years, journalists have rather optimistically covered new research indicating some lab-engineered enzymes and bacteria can break plastic down into its essential molecular components. This would hypothetically improve the quality of some types of recycled plastic, most notably PET, which is commonly used to make plastic bottles. This, though humanity does not yet possess the recycling infrastructure or technology to employ such "plastic-eating" enzymes and bacteria on a useful scale. When pressed to put their findings into context, experts admit future research will be required to determine if doing such a thing is even feasible.[28]

The bottom line is that, even with bio-engineered "plastic eaters," recyclers will still have to collect, clean, and process used plastic, on top of continuing efforts to educate people on how to direct their plastic waste to recycling facilities. While some scientists say it's a matter of time before we find plastic-digesting microbes in the environment—the result of evolution—we wouldn't necessarily want plastic-hungry organisms to take over the environment.[29] You could imagine what would happen if such a bacteria or enzyme began eating away at the plastic lines carrying gasoline in your car, or the plastic-wrapped foods in a grocery store, for example. The unintended results of a mass dissolution of all the plastic presently on Earth could prove both wonderful and disastrous—for humans, and for all of nature that's rapidly trying to adjust to living among our waste.

With all due respect to the potential benefits of ongoing and future scientific research and discovery, it seems one of the greatest lessons to learn from plastic is to first slow down, be present, and understand the full range of consequences of invention before worshipping any material reminiscent of a quick fix.

# CHAPTER 12
# Circular Thinking

Nearly four years out from our expedition across the eastern North Pacific Gyre, Malene Møhl was hired by the city of Copenhagen as a plastic consultant. She was doing so at an exciting time, as the EU had since passed a directive requiring member states to eliminate some of the most commonly littered single-use plastic items, including balloons and plastic bags, among other measures to reduce waste and help facilitate a circular economy.[1] A completely circular, global system of plastic use, collection, and reuse is still a dream of the future. But some places are experimenting with schemes like this on a small scale—with hopes of increasing in scope and ambition as they learn. In Denmark, Malene has contributed to one such initiative.

Denmark wasted no time in formulating an action plan outlining how it would adhere to the new EU legislation. By the summer of 2018, municipal employees in the capital had launched an initiative called "Circular Copenhagen," which Malene joined in 2020. One of its primary goals included establishing local waste collection programs and infrastructure capable of recycling 70 percent of the city's municipal trash, plastic included, by 2024.[2]

This would be a big shift, even by Copenhagen standards. Across Denmark, about 60 percent of all collected plastic waste, along with much other trash, is burned.[3] Copenhagen's incineration complex, the Amager Resource Center, emerges like a steep, shining mountain from Denmark's otherwise pancake-flat island of Amager. Its designers, a team of architects at Bjarke Ingels Group (BIG), plunked a few peculiar features on the aluminum-wrapped power plant, a quarter-mile-long public ski slope and an outdoor climbing wall among them. Since the $600 million-plus plant was brought online in 2017, hundreds of trucks carrying waste collected off Copenhagen's streets have dumped their hauls here daily. To keep the facility's extra-large furnaces full, the Amager Resource Center also regularly receives trash exported from elsewhere in Europe. Heat generated by the furnaces is harnessed to meet local needs for indoor heating and electricity.[4]

Of all countries in Europe, the Nordic nations—save for Iceland—are among the most reliant on incineration as a means of waste management. For decades, this region touted its lack of landfills as evidence it was expertly handling its residents' waste. The reality is that Scandinavia has been burning much of its garbage—plastic and otherwise—and passing that bulk of incinerated trash off as being recycled.[5] In 2018, Denmark adopted a strategy requiring that 80 percent of all plastic waste historically sent to incinerators be recycled instead by 2030. This was done to minimize emissions of harmful climate-warming gases discharged during incineration, and to maximize the recovery of recyclable and potentially valuable resources. However, at the time Malene was hired, necessary improvements in collection, sorting, and processing were still in development—and so much of the city's plastic and other trash would continue to be burned instead of recycled.[6]

"As we've seen so many times before, bureaucratic contradictions prevent necessary action," Malene observed.

By the time she assumed her new post, Copenhagen was still oceans away from meeting the EU's new standards for recycling plastic waste, and from reaching its own ambitions of becoming the world's first carbon-neutral city by 2025.[7] Fulfilling these objectives would require Copenhagen to seriously shift the way it handled its trash, particularly plastic packaging—a ubiquitous element of modern life across Scandinavia, and indeed so much of the planet.

"In a truly 'circular economy,' a material can become itself again after its final use, and remain the same quality," said Malene, adding that was without the addition of freshly made plastic or toxic plasticizers. "Unfortunately, a lot of the plastic packaging we make and use is very difficult to recycle."

Polyethylene terephthalate, better known as PET, or PETE, is an exception. You can tell an item is made from PET, a type of plastic related to polyester, if it's marked with the resin code number 1. PET is unusually well suited for repeated recycling because it can be melted down at a relatively low temperature while maintaining its useful qualities: moisture resistance, strength, and light weight among them. Most other plastics must be melted down at extremely high temperatures when recycled, and in the process, these plastics tend to lose their marketable traits. This renders them unsuitable for a return to their original form after recycling, or, in order to return, requires the addition of generous amounts of freshly made plastic, hazardous solvents, hormone-disrupting plasticizer chemicals, and other additives.[8] PET, being less absorbent than other kinds of plastic, does not tend to pick up grease, soaps, and other residues that often contaminate potentially recyclable plastic waste. This makes it relatively simple to sanitize, particularly after use as packaging for food and beverages, for which it is widely employed.

PET also happens to be highly abundant in Copenhagen's waste stream: More than one-third of the plastic waste produced by the city's residents is food trays and tubs, many made of PET and others made of a less expensive type of plastic called polypropylene (PP). PP is more challenging to recycle, and to do so requires the input of fresh plastic and chemicals. Until now, much of the city's PET food packaging—like all the other plastic items it collects—have been either recycled for nonfood purposes or incinerated, as is the status quo. HDPE, or high-density polyethylene, commonly used to make hard plastic food and beverage containers, also seems to recycle better than other plastics—with little to no need for additives.[9]

If Denmark recycled all the plastic waste it presently burns, the country could reap an economic benefit of 1.6 billion Danish kroner (more than $250 million) annually and create a significant number of high-income jobs, according to marketing research firm McKinsey and Company. And incineration only drives the need for more plastic production: Lacking a steady stream of high-quality recycled plastic, the food industry here and globally is a prolific purchaser of the freshly made, never-before-recycled stuff.[10] The industry's demand and an absence of circularity drives plastic manufacturers to continue producing petro-chemical-based plastic.

During the spring of 2020, while Denmark kept its borders sealed amid the COVID-19 pandemic, Malene and her colleagues quietly embarked on a pilot project they hoped could make a case for plastic circularity. They worked with the city's recyclers to send about eight metric tons of PET food trays skimmed out of the city's waste stream to a Danish-owned recycling company, 4PET, in Duiven, Netherlands. In the Netherlands, the collected plastic food trays were sanitized and melted down into food-grade PET pellets. The PET pellets were shipped back to Denmark, where Færch Plast, a food-packaging manufacturer that now owns 4PET, molded the pellets back into about 250,000 food trays.[11] There's a way to go before the process can be deemed completely circular, as the trays do contain 10 percent fresh plastic. But Copenhagen's meat trays still contain significantly less added plastic than most other food-grade plastic products on the market today.[12]

Færch Plast is one of a small but growing number of companies across various economic sectors that are beginning to recognize the potential for a public image boost in the eyes of a society increasingly concerned about plastic and the even more time-sensitive issue of climate change. Pushed by newly passed legislation, and armed with the recent scientific knowledge that PET—as well as HDPE—can be recycled repeatedly with little to no inputs of freshly made plastic or additional chemicals, some corporations are now experimenting with efforts to create products and packaging either partly or entirely from recycled plastic. These include branches of some major brands that, being in the business of selling unfathomable amounts of plastic-wrapped products, are notorious contributors to the world's plastic pollution load. Beverage giants Coca-Cola and PepsiCo, and prepackaged-snack empire Mondelez, are among them.[13]

European corporations Hilton Foods and Danish Crown packed the fully recycled plastic food trays with chicken, beef, and Danish pork. Meat carried in the recycled PET packages, which were prominently labeled as such, was sold in Danish supermarkets Coop and REMA 1000 throughout the summer.

According to Malene, Danes weren't repulsed by eating food sold in material pulled from packages once piled up as garbage. In fact, she said, it seemed many customers were more inclined to purchase meat wrapped in the recycled trays over meat wrapped in new plastic trays, a tiny testament to the capacity of human values to shift toward circularity—the only system nature has proven to last, over and over again.

"It's absolutely necessary to cycle all the materials people use, including plastic, like the natural environment does with minerals and atoms," Malene told me.

When Copenhagen's meat-tray recycling experiment worked, turning all collected meat trays back into new meat trays, without creating waste, at a cost palatable to a plastic manufacturer, recycler, corporation, government, and consumer, it became one of the first small success stories making a case for circularity on a municipal level. Yet recycling infrastructure and collection systems tailored to plastic, and the circular values needed to support a new forward-thinking economy, are still nascent globally. This, despite rising awareness of the problems plastic pollution brings and increasing public demand for recycled products and packaging. Without comprehensive legislation to regulate plastic production and use, there's little incentive for the plastic industry to participate in a recycling revolution.

As we've seen, up to the present day, a combination of local and sometimes national regulations on plastic production, sale, use, disposal, and recycling—or complete lack of regulation—sway the way we handle plastic waste across place and context. Primarily, municipal and national governments have tried to encourage shifts in values and behaviors through disincentivizing the purchase or distribution of single-use plastic items, through implementation of taxes and in some cases prohibition. It's only recently that some have proposed truly circular plastic legislation, much of it requiring the companies that create products made from plastic to finally take full responsibility for their waste.

A shift is happening, on a large scale, as a growing number of lawmakers voice support for circular policies. The European Commission's directive on single-use plastic is considered one of the biggest steps taken to treat plastic and other materials people use in a less-wasteful way. Its nuanced policy for curbing plastic pollution is based around a simple premise: Making better use of the plastic we already have reduces the need to produce more plastic. Among the strategies outlined in this and other circular plans are obligations for corporations to assume extended producer responsibility, or EPR for short, assuming economic and ecological accountability for their products throughout the products' entire life cycles. EPR schemes may involve continued research, deposit-return systems, vastly improved recycling systems, redesign and replacement, use restrictions, and better collection systems for many popular single-use plastic products, among other initiatives.[14]

Unlike most plastic legislation passed to date, which has done little to shift throwaway culture, the EU's directive seeks to rethink plastic as a resource, instead of waste, and close the present gap that exists between a plastic item's final use and its potential next life—which is rarely ever realized.

For all the people on board with circularity, there are many others who are not—unsurprisingly, most people leading industries dealing in petrochemicals and plastic. While some companies, including Coca-Cola, have publicly committed to some voluntary measures to cut plastic use, behind closed doors it and other companies have opposed important policies that address plastic, including the EU directive. Further adding to the illusion of corporate concern is the common practice of allowing industry trade associations to do their dirty work.

"While companies may tell the public they're good guys, many belong to trade groups—including the Plastics Industry Association—which are strongly opposing the legislation we need right now," said John Hocevar, of Greenpeace, who is now working to hold some of the world's biggest companies accountable for their secretive trade association alliances. "We call up these companies and remind them that they are deceiving people by saying publicly that they care, but in reality belong to groups opposing meaningful action."

As a result of such efforts, big brands like Coca-Cola, General Motors, PepsiCo, and SC Johnson severed ties with the Plastics Industry Association in 2019. Still, when plans for the EU directive were first unveiled, Coca-Cola (then still a member of the trade group), was the biggest corporation to sign off on a letter to the European Commission opposing the new plans, which would require that manufacturers redesign their plastic bottles so that the caps were less likely to twist off and become unrecoverable in nature, among other measures meant to minimize plastic products' harm on the natural environment and improve plastic recovery and recycling rates.

In their letter, the beverage corporation leaders cite the efficacy of deposit return schemes and recycling in reducing plastic litter in their arguments against the EU directive, which would require serious commitment and investment by corporations. This, though Europe's average plastic recycling rate, while higher than in many parts of the world, is nowhere near circular at just 42 percent, with much of it exported elsewhere, to be burned or piled up in landfills and the natural environment instead of actually being recycled.[15] The corporations proposed increased efforts to "reinforce and incentivize [the] right consumer behaviors" in lieu of taking responsibility for their products.[16] It's the same old story.

When politicians in the world's most wasteful country, the US, unveiled its first national circularity-based plan to tackle plastic pollution in 2020, the plastic industry, predictably, reacted in strong opposition. Called the Break Free From Plastic Pollution Act, and recently reintroduced in March 2021 by Congressman Alan Lowenthal (D-CA), Senator Jeff Merkley (D-OR), and more than ninety other members of the House and Senate, the act is designed to ultimately compel corporations and industries to cease production of certain non-recyclable single-use plastic products. To achieve this goal, the act would, among other strategies, require governments and industries to assume additional responsibility for plastic products, phase out some single-use plastic products entirely, restrict plastic waste exports, and place a temporary moratorium on permits for new and expanded plastic- and petrochemical-producing facilities.[17]

Judith Enck, president of Beyond Plastics, an organization focused on eliminating plastic pollution, has proven to be a valuable ally of the Break Free From Plastic Pollution Act, advising lawmakers and other supporters of the bill. As a former EPA administrator, she is well aware of both the government bureaucracy and industry influence that so commonly impede passage of meaningful legislation. "The biggest hurdle to getting the Break Free From Plastic Pollution Act passed is to get past plastic lobbyists in every state legislature and Congress who are trying to tinker with the language of the law so that it's less effective," Enck said in early 2021, the bill still under consideration.

When asked directly about the issue of plastic pollution and how to best address it, a representative from the Plastics Industry Association, the major plastic-industry trade group, told me in an email that it "believes uncollected plastics do not belong in the natural environment and that is why we partner with other associations, non-governmental organizations, and intergovernmental authorities to coordinate efforts to strengthen recovery systems around the globe to prevent loss of plastics into the environment. Our members understand that our industry needs to be a part of the solution. We encourage education and call for the enhancement of our recycling infrastructure in order to encourage new end markets for plastics."

Scientists continue to reiterate that industry's inclination to put the onus for plastic pollution almost solely on consumers is unfair.

"Ocean plastics are a symptom of poor upstream waste management, poor product design, as well as consumer littering behavior," Marcus Eriksen of 5 Gyres once explained to me. What industry suggests as a solution is "a perpetuation of old narratives, where pollution is caused by consumers. Regulation of products and packaging must be fought for intensively."

Those laws now in place have already proven themselves on varying scales over the past few decades. Local rules on single-use plastic have been linked to reduced amounts of plastic waste ending up as litter in the environment.[18] But strong national legislation—and hopefully, one day international legislation—is by far most capable of making the biggest reductions in plastic production, use, and disposal, due to the global nature of the plastic crisis.

To get effective legislation passed even in the face of widespread industry opposition, Enck urges people to contact their elected officials and express why we need strict plastic legislation now. "Yes, it's hard, and there's a lot already on people's plates," she acknowledged. "But believe it or not, many lawmakers are still fixed on the idea that plastic pollution is 'just' a straw up a turtle's nose." She suggested the public remind their representatives that the plastic crisis is much larger and more urgent than that single perspective—causing not only ecological catastrophes but also harming human health, while upholding systemic racism and other forms of injustice.

And then there is plastic's inherent connection to fossil fuels, and the catastrophe that is climate change. Because fossil fuels are finite—their underground stores cannot be replenished when exhausted—there will inevitably come a day when petrochemical industries and manufacturers will have to rethink their reliance on plastic. Looking forward, it seems most players in industry and business will remain focused on perpetuating the plastic status quo, all while continuing to rake in billions of dollars a year—at great expense to all of us. That is, of course, unless we stop them.

Is it brilliant, or brainless, that petrochemical and plastic corporations are choosing to go down with a sinking ship?

The answer to that question depends on what matters most to each of us.

CONCLUSION
# Giants Do Fall

*Plastic* as a word is pulled from the Latin *plasticus* and Greek *plastikos*, terms used to indicate a material's ability to be molded or shaped.[1] It's this single quality that makes plastic as we know it—the fossil-fuel derived "material of a thousand uses," as Leo Baekeland first sold it—so extraordinary.[2]

Lest the consuming public forget that fact, the American Chemistry Council has chanted the mantra "Plastics Make It Possible" in print, over the airwaves, across the web, and on TV for decades.[3] Plastic can be shaped into shopping bags or car airbags; shirts or medical syringes; soda bottles or bandages; packaging or pacemakers; chewing gum or gun parts. Plastic shapes human identity and speeds up the rate at which we move across the world and through our days, connecting people and allowing us to express who we are to each other. And yet plastic also helps us destroy. Plastic has saved our lives, while taking others' away. Plastic is a miracle. Plastic is a scourge.

As people have learned from industries how to use plastic, our culture has been shaped too. We have become plastic people of the plastic age. Plastic is engrained in our bodies, our identities, our values, our common home.

We plastic people have a far reach. Cutting fast above our heads at all times are bits of plastic debris creating a minefield woven in the cosmos. There are hundreds of thousands of manmade objects and fragments, mostly metals and plastics, larger than one centimeter in diameter now orbiting Earth, with hundreds of millions smaller than one millimeter, and a hundred trillion smaller than one micron.[4] Humanity's plastic detritus has not only come to dominate Earth and its inhabitants; it now has the whole planet surrounded.

Each ad we hear or see is expertly crafted to rapidly convince us, in a beat, a glance or two, that what is being sold is what we need *now*. What happens to these items in a year, a month, even the next five minutes, and why we feel compelled to buy it, may not even emerge as the faintest thoughts in our minds. When our stuff wears out or goes out of fashion, we buy new stuff. This "planned obsolescence" has become an intentional business strategy.[5]

"Much of our waste problem is to be accounted for by the intentional flimsiness and unrepairability of the labor-savers and gadgets that we have become addicted to," wrote Kentucky-raised author and farmer Wendell Berry in his 1990 book of ruminations on humanity, *What Are People For?* "The truth," Berry concludes, "is that we Americans, all of us, have become a kind of human trash, living our lives in the midst of a ubiquitous damned mess of which we are at once the victims and the perpetrators."[6]

Of course, while materialism is infused with a particular brand of American ethos, it's not just Americans who are afflicted by the object-laden wealth that's come to influence our understanding of what it means to live a good life. Though materialism doesn't necessarily bother all people as much as it irks Berry, the undeniable fact is that the production, use, and disposal of plastic is linked to numerous ecological and social disasters. We were never brainwashed, only bribed, to covet the things that don't matter. And in the process, we're destroying the only things that do.

Having heard no word on their appeal or lawsuit by midsummer 2020, RISE and its allies again took Formosa to federal court, this time to file an injunction to halt construction activities. In an ongoing lawsuit, the plaintiffs alleged the US Army Corps of Engineers did not sufficiently evaluate the Sunshine Project's expected impacts on the unmarked graves and air pollution, nor did it sufficiently review how constructing a massive petrochemical complex across a naturally protective wetland area might contribute to flooding and reduced hurricane resistance in the future.[7] The Corps' stated mission is to "deliver vital public and military engineering services; partnering in peace and war to strengthen our Nation's security, energize the economy and reduce risks from disasters."[8]

Finally, Formosa Group flinched. The company agreed to halt all major construction activities until February 1, 2021, specifically the creation of a loading dock on the Mississippi River, development of the site's ecologically valuable wetlands, and the disruption of all suspected gravesites, including Buena Vista cemetery. In exchange, RISE and its allies dropped their request for a preliminary injunction on the lawsuit. Under the present agreement, Formosa is permitted to continue limited pre-construction activities, such as widening Highway 3127, relocating utilities, and testing soil.[9]

While Formosa representatives publicly shrugged off the agreement as a mere inconvenience, Sharon Lavigne and her neighbors remained cautiously optimistic that the extra time the agreement afforded would give them a key opportunity to strategize; they would find a way to stop Formosa once and for all. They'd move forward fighting their legal battles with Formosa with or without the help of their state's elected officials, who, up to that point, showed no interest in RISE's cause. And by taking action to address the plastic crisis and the toxic systemic issues at its core, Lavigne's community has managed to capture the world's attention—at a critical moment—and the world is rapidly responding to their calls for support. United Nations human rights experts have recently condemned Formosa's Sunshine Project and the continued industrialization of Cancer Alley as environmental racism and called on the US government "to deliver environmental justice in communities all across America, starting with St. James Parish."[10]

As COVID-19 raged on through what remained of 2020, an abnormally warm ocean spun several major tropical storms, cyclones, and hurricanes, which pummeled islands and coastal areas; deadly heat waves with temperatures near 130 degrees Fahrenheit scorched much of the Middle East; and record-high heat and dry conditions on the American West Coast, throughout the Amazon, and across Siberia fueled immense and destructive forest fires.[11] The year 2020 proved that exploiting fossil fuels has already rendered significant areas of the world unlivable for people and much other life.[12] People with the means are relocating, while many more lacking resources to head for more habitable ground continue to suffer. Many of us have been fortunate enough to be spared from either of these fates—for now. But our own difficult, dismal futures are not far off if we do not take decidedly swift and significant action to curb carbon emissions.[13]

If fitted together, all oil and gas pipelines presently laid could wrap around the planet more than fifty times.[14] In most cases, this rapidly expanding petrochemical infrastructure is buried beneath our feet, marked aboveground by minuscule signs softly warning "pipeline," when marked at all. For most of us, the oil and gas infrastructure underlying our everyday life exists out of sight, out of mind—away. But this is not so for the people of Cancer Alley, the Gulf Coast, Appalachia, the Ohio River Valley, and other industrial regions far beyond.

Here, the consequences of a polluted world and a consequently unstable climate are glaringly apparent. In standing up for themselves, people living on the fenceline are showing all of us a different, better future is possible. Change is happening, now.

"Giants do fall," said Stephanie Cooper. "Money is powerful, but not more powerful than human life."

# Acknowledgments

Thank you to all who bravely bear witness, speak out, and take action on the front lines of the plastic crisis, for you have made this book possible. I am deeply grateful to the many people who have shared their testimony and time with me.

Particular thanks to the sailors with whom I journeyed far and wide —and, along the way, became something like family to me—aboard *Christianshavn*, *Ópal*, and *TravelEdge*; and of course to our captains Torsten Geertz, Heimir Harðarson, and Anna Strang; to first mates Rasmus Hytting and Maggie Kerr; and to deckhand Sophie Dingwall—for staying the course, and keeping us afloat.

For showing me the depths of the plastic crisis, I thank the many scientists who kindly invited me to spend numerous hours with them in their offices, laboratories, and in the field—you are some of the most incredible teachers I know: Kristian Syberg, Malene Møhl, Alvise Vianello, Jes Vollertsen, Sam Mason, Joseph Gardella Jr., Courtney Wigdahl-Perry, Marcus Eriksen, Charlie Moore, Megan Lamson, Ruth Gates, Craig Colten, Winnie Courtene-Jones, Charla Basran, and Daniel Gonzalez de la Peña. I also thank the other experts who fielded my endless questions about plastic in conversations and other correspondence: Rebecca Altman, Heather Barrett, Joana Correia Prata, Abigail Snyder, Jenna Jambeck, Chelsea Rochman, Kara Lavender Law, and Sedat Gündoğdu, among them. I am deeply grateful to Roskilde University, which provided me with the space and resources necessary to do much vital research for this project.

Thank you to the brave individuals and communities on the front lines of the plastic crisis for showing me (and all of us) how positive change is made: Sharon Lavigne and RISE St. James, Anne Rolfes and Louisiana Bucket Brigade, Bill Gilmartin and Hawai'i Wildlife Fund, Renate Heurich and 350 New Orleans, Catherine Comeaux and No Waste Nola, Rafael Bergstrom and Sustainable Coastlines Hawai'i (formerly Surfrider Foundation, O'ahu Chapter), Stuart Coleman and Wastewater Alternatives and Innovations (also formerly Surfrider Foundation, O'ahu Chapter), Emily Penn and eXXpedition, Delia Creamer and the Center for Biological Diversity, Rich Cramp and Trash Hero, John Hocevar and Greenpeace, Krix Luther and Clean the Beach Boot Camp, Henrik Beha Pedersen and Plastic Change, Lisbeth Engbo and Ekspedition Plastik (formerly Plastic Change), Bélen Garcia Ovide and Ocean Missions, Peggy Shepard and WE ACT for Environmental Justice, Rasmus Holm and Skraldejagt, Judith Enck and Beyond Plastics, Sindy Yilmaz, Malcolm Thompson and Save Our Skibbereen. I also appreciate the incredible artists who spent considerable time showing and explaining to me how they are pioneering use of a new creative media—plastic: Swaantje Güntzel, Cindy Pease Roe, Pam Longobardi, Kristian Brevik, and Chris Jordan, among them. Thanks too to the innovators who have discussed with me their groundbreaking efforts to eliminate plastic waste from the things we buy and the environment we live in: Cyrill Gutsch and Parley for the Oceans, Boyan Slat and the Ocean Cleanup, Andrew Turton and Pete Ceglinski of the Seabin Project, John Kellett and Clearwater Mills, Joey Zwillinger and Allbirds, Brian Bushell and *by* Humankind, Natsai Audrey Chieza and Faber Futures, Laura Luchtman and Kukka and the Living Colour Collective, Eben Bayer and Gavin McIntyre of Ecovative, Theanne Schiros of FIT and Werewool, and many others.

I am grateful for fellowships awarded by the Safina Center, Woods Hole Oceanographic Institution, and Craig Newmark Graduate School of Journalism at CUNY, and generous gifts offered by friends and family—which helped cover some of the costs of my numerous research trips. I also extend my gratitude to the editors who have published my writing and images illuminating the plastic crisis in print and online, including Zenobia Jeffries Warfield, John Platt, Colin Schultz, Todd Woody, Dean Visser, Tanya Lewis, Derek Mead, Elizabeth Limbach, and Purbita Saha—and of course, a heartfelt thanks to my amazing team at Island Press: Rebecca Bright, the talented editor of this book; Sharis Simonian, dedicated production manager; and Elizabeth Farry, meticulous copyeditor. And thank you to my fabulous agent, Ariana Philips, for believing in this book—and that I could write it.

Thank you to all my friends, especially Danielle ("Danzo") DeLaurentis and Jen Chiodo, who not only cheered me on but also cheered me up when the often-heavy subject matter of this book began weighing me down. ("Don't be stressed, be blessed." I won't soon forget!) Thanks to Lise Hintze (and Polly Jane Hintze) for all the support, kindness, and puppy playdates. For inspiring me to start writing about our precious and imperiled planet in the first place, I thank Heidi Hutner—my friend, colleague, and the best college professor I ever had. Much gratitude to family: my mother and brother, Laura and Evan Cirino, and Danielle Messina, for all they have done to support and encourage me, and for taking good care of our late beloved Foosa, legendary Alaskan malamute, while I was out at sea. And a warm thanks to Sabi, the special dog who made the experience of finishing a book during a global pandemic far less lonely than it otherwise could have been.

Lastly, special thanks to my writing role model, friend, and colleague, Carl Safina, whose suggestion to sail into the Garbage Patch became the inspiration for the book that is now in your hands—and whose mentorship and generosity continue to bring me to new and exciting places.

# Notes

## Foreword

1. Ammer, Christine. 2013. *The Dictionary of Clichés*. New York: Simon & Schuster. Accessed via the Farlex Free Dictionary. https://idioms.thefreedictionary.com /waste+not+want+not.

## Chapter 1

1. National Geographic Resource Library. 2019. "Great Pacific Garbage Patch." https://www.nationalgeographic.org/encyclopedia/great-pacific-garbage-patch/.
2. Pedersen, H. B. 2016. In-person interview.
3. Plastic Change. 2019. "Ekspedition Plastic Change (2016–2018)." https:// plasticchange.dk/videnscenter/ekspedition-plastik/.
4. National Oceanic and Atmospheric Administration. 2019. "How Big Is the Great Pacific Garbage Patch? Science vs. Myth." Office of Response and Restoration. https://response.restoration.noaa.gov/about/media/how-big-great-pacific -garbage-patch-science-vs-myth.html; Syberg, K. 2016. In-person interview.
5. National Oceanic and Atmospheric Administration. 2021. "What Is Ghost Fishing?" National Ocean Service. https://oceanservice.noaa.gov/facts/ghost fishing.html.
6. Møhl, M. 2016 and 2020. In-person interviews.
7. Dolman, S. J., and P. Brakes. 2018. "Sustainable Fisheries Management and the Welfare of Bycaught and Entangled Cetaceans." *Frontiers in Veterinary Science* 5 (287). https://doi.org/10.3389/fvets.2018.00287.
8. Thomas, S. N., and K. M. Sandhya. 2020. "Netting Materials for Fishing Gear with Special Reference to Resource Conservation and Energy Saving." *ICAR Research Data Repository for Knowledge Management.* https://krishi.icar.gov.in

/jspui/bitstream/123456789/30998/2/03_Netting%20Materials%20for%20 Fishing%20Gear.pdf.

9. Richardson, K., et al. 2019. "Estimates of Fishing Gear Loss Rates at a Global Scale: Literature Review and Meta-Analysis." *Fish and Fisheries* 20 (6).

10. Baheti, P. 2021. "How Is Plastic Made? A Simple Step-By-Step Explanation." British Plastic Federation. https://www.bpf.co.uk/plastipedia/how-is-plastic -made.aspx.

11. Ritchie, H., and M. Roser. 2017. "Fossil Fuels." Our World In Data. https:// ourworldindata.org/fossil-fuels.

12. York, R. 2017. "Why Petroleum Did Not Save the Whales." *Socius*. https://doi .org/10.1177/2378023117739217.

13. National Geographic Resource Library. 2013. "The History of the Ivory Trade." https://www.nationalgeographic.org/media/history-ivory-trade/.

14. Boyd, J. E. 2011. "Celluloid: The Eternal Substitute." Science History Institute. https://www.sciencehistory.org/distillations/celluloid-the-eternal-substitute.

15. Buchanan, R. A., et al. 2020. "History of Technology: Plastic." *Encyclopedia Britannica.* https://www.britannica.com/technology/history-of-technology /Atomic-power#ref14890; Boon, R. 2014. "Alexander Parkes: Living in a Material World." *Science Museum.* https://blog.sciencemuseum.org.uk/alexander -parkes-living-in-a-material-world/.

16. "Hyatt Celluloid Billiard Ball." National Museum of American History Behring Center. Smithsonian Institute. https://americanhistory.si.edu/collections/search /object/nmah_2947.

17. Boyd. "Celluloid."

18. American Chemical Society. 1993. "Leo Hendrik Baekeland and the invention of Bakelite." https://www.acs.org/content/acs/en/education/whatischemistry /landmarks/bakelite.html.

19. "Leo Baekeland Diary Volume 01, 1907–1908." Transcribed and Reviewed by Digital Volunteers Approved by Smithsonian Staff. Retrieved April 8, 2020, https://edan.si.edu/transcription/pdf_files/6607.pdf.

20. American Chemical Society. "Leo Hendrik Baekeland and the invention of Bakelite."

21. Science History Institute. 2017. "Leo Hendrik Baekeland." https://www .sciencehistory.org/historical-profile/leo-hendrik-baekeland.

22. Syberg, K. In-person interview.

23. Whiting, K. 2018. "This Is How Long Everyday Plastic Items Last in the Ocean." World Economic Forum. https://www.weforum.org/agenda/2018/11 /chart-of-the-day-this-is-how-long-everyday-plastic-items-last-in-the-ocean/.

24. Andrady, A. L. 2015. "Persistence of Plastic Litter in the Oceans." *Marine Anthropogenic Litter*, 57–72. https://doi.org/10.1007/978-3-319-16510-3_3.

## Chapter 2

1. Turns, A. 2018. "Saving the Albatross: 'The War Is Against Plastic and They Are Casualties on the Frontline.'" *Guardian*. https://www.theguardian.com /environment/2018/mar/12/albatross-film-dead-chicks-plastic-saving-birds; Jordan, C. 2017. *Albatross*. https://www.albatrossthefilm.com/.

2. Thompson, R. C. 2004. "Lost at Sea: Where Is All the Plastic?" *Science* 304 (5672): 838 https://doi.org/10.1126/science.1094559; Syberg, K. 2016. In-person interview.

3. Koelmans, A. A., E. Besseling, W. J. Shim. 2015. "Nanoplastics in the Aquatic Environment. Critical Review." In Bergmann, M., L. Gutow, and M. Klages (eds), *Marine Anthropogenic Litter*. Springer, Cham, 325–40. https://doi.org /10.1007/978-3-319-16510-3_12.

4. British Plastics Federation. 2021. "Plastic Applications." https://www.bpf.co.uk /plastipedia/applications/default.aspx.

5. Parker, L. 2018. "Facts About Plastic Pollution." *National Geographic.* https:// www.nationalgeographic.com/science/article/plastics-facts-infographics-ocean -pollution.

6. UNEP. 2018. "Our Planet Is Drowning in Plastic Pollution—It's Time for Change!" https://www.unep.org/interactive/beat-plastic-pollution/.

7. Geyer, R., et al. 2017. "Production, Use, and Fate of All Plastics Ever Made." *Science Advances* 3 (7). https://doi.org/10.1126/sciadv.1700782.

8. Reis, F., et al. 2019. "Life in Plastic, It's Not Fantastic: The Economics of Plastic Pollution." *Science for Sustainability Journal* 3. https://www.greenofficevu.nl /wp-content/uploads/2019/08/Life-In-Plastic.pdf.

9. Geyer et al. "Production, Use, and Fate."

10. Geyer et al. "Production, Use, and Fate."

11. Pedersen, H. B. 2016. In-person interview.

12. Syberg, K. 2016. In-person interview.

13. Liboiron, M. 2015. "BabyLegs." CLEAR. https://civiclaboratory.nl/2015/05 /31/babylegs/.

14. Møhl, M. 2016. In-person interview.

15. National Ocean Service. 2009. "How Much of the Ocean Have We Explored?" National Oceanic and Atmospheric Administration. https://oceanservice.noaa .gov/facts/exploration.html.

16. United Nations. 2017. "'Turn the Tide on Plastic' Urges UN, as Microplastics in the Seas Now Outnumber Stars in Our Galaxy." UN News. https://news .un.org/en/story/2017/02/552052-turn-tide-plastic-urges-un-microplastics -seas-now-outnumber-stars-our-galaxy.

17. Gigault, J., et al. 2018. "Current Opinion: What Is a Nanoplastic?" *Environmental Pollution* 235 (April): 1030–34. https://doi.org/10.1016/j.envpol.2018.01.024.

18. Syberg, K. In-person interview.
19. Syberg, K. In-person interview.
20. Ocean Conservancy and McKinsey Center for Business and Environment. 2015. "Stemming the Tide: Land-Based Strategies for a Plastic-Free Ocean." https://oceanconservancy.org/wp-content/uploads/2017/04/full-report -stemming-the.pdf.
21. Eriksen, M., et al. 2014. "Plastic Pollution in the World's Oceans: More Than 5 Trillion Plastic Pieces Weighing Over 250,000 Tons Afloat at Sea. *PLoS ONE* 9 (12): e111913. https://doi.org/10.1371/journal.pone.0111913.

## Chapter 3

1. Hansen, A. 2011. "Glowing Plankton. ASU—Ask A Biologist." https://aska biologist.asu.edu/glow-dark-plankton.
2. Møhl, M. 2016. In-person interview.
3. Syberg, K. 2016–2020. In-person interviews.
4. The Great Nurdle Hunt. n.d. "The Problem." https://www.nurdlehunt.org.uk /the-problem.html.
5. Food and Agriculture Organization of the United Nations. 2018. *The State of World Fisheries and Aquaculture: Meeting the Sustainable Development Goals.* The State of the World series. Rome, Italy: FAO. http://www.fao.org/3/i9540en /i9540en.pdf.
6. Food and Agriculture Organization of the United Nations. 2019. *The State of Food Security and Nutrition in the World: Safeguarding Against Economic Slow-downs and Downturns.* The State of the World series. Rome, Italy: FAO. http:// www.fao.org/3/ca5162en/ca5162en.pdf; United States Census Bureau. 2021. "U.S. and World Population Clock." https://www.census.gov/popclock/.
7. Kourous, G. 2005. "Many of the World's Poorest People Depend on Fish." Food and Agriculture Organization of the United Nations. http://www.fao.org /newsroom/en/news/2005/102911/index.html.
8. Drake, D. C., et al. 2006. "Fate of Nitrogen in Riparian Forest Soils and Trees: AN15N Tracer Study Simulating Salmon Decay." *Ecology* 87 (5): 1256–66. https://doi.org/10.1890/0012-9658(2006)87[1256:fonirf]2.0.co;2.
9. Gulick, A. 2010. *Salmon in the Trees: Life in Alaska's Tongass Rain Forest.* Seattle, WA: Braided River.
10. Ellen MacArthur Foundation. 2016. "The New Plastics Economy: Rethinking the Future of Plastics and Catalysing Action." https://www.ellenmacarthur foundation.org/publications/the-new-plastics-economy-rethinking-the-future -of-plastics-catalysing-action.
11. Elhacham, E., et al. 2020. "Global Human-Made Mass Exceeds All Living Bio-mass." *Nature* (December): 1–3. https://doi.org/10.1038/s41586-020-3010-5.

12. Miranda, D. A., and G. F. de Carvalho-Souza. 2016. "Are We Eating Plastic-Ingesting Fish?" *Marine Pollution Bulletin* 103 (1): 109–14. https://doi.org/10.1016/j.marpolbul.2015.12.035.

13. Møhl, M. In-person interview.

14. Savoca, M. 2018. "The Ecology of an Olfactory Trap." *Science* 362 (6417): 904. https://doi.org/10.1126/science.aav6873.

15. Wilcox, C., et al. 2015. "Threat of Plastic Pollution to Seabirds Is Global, Pervasive, and Increasing." *Proceedings of the National Academy of Sciences* 112 (38): 11899–904. https://doi.org/10.1073/pnas.1502108112.

16. Pew Trusts. 2013. "Forage Fish." https://www.pewtrusts.org/en/research-and-analysis/fact-sheets/2013/09/25/forage-fish-faq.

17. Savoca, M. S., et al. 2017. "Odours from Marine Plastic Debris Induce Food Search Behaviours in a Forage Fish." *Proceedings of the Royal Society B: Biological Sciences* 284 (1860): 20171000. https://doi.org/10.1098/rspb.2017.1000.

18. Allen, A. S., et al. 2017. "Chemoreception Drives Plastic Consumption in a Hard Coral." *Marine Pollution Bulletin* 124 (1): 198–205. https://doi.org/10.1016/j.marpolbul.2017.07.030.

19. Reichert, J., et al. 2018. "Responses of Reef Building Corals to Microplastic Exposure." *Environmental Pollution* 237 (June): 955–60. https://doi.org/10.1016/j.envpol.2017.11.006.

20. Cesar, H., et al. 2003. "The Economics of Worldwide Coral Reef Degradation." Cesar Environmental Economics Consulting. https://wwfint.awsassets.panda.org/downloads/cesardegradationreport100203.pdf.

21. National Oceanic and Atmospheric Administration. 2019. "Polyps Up Close." National Ocean Service. https://oceanservice.noaa.gov/education/tutorial_corals/media/supp_coral02bc.html; Florida Keys National Marine Sanctuary. n.d. "Corals Get Their Food from Algae Living in Their Tissues or by Catching and Digesting Their Prey." National Oceanic and Atmospheric Administration. https://floridakeys.noaa.gov/corals/coralseat.html.

22. National Oceanic and Atmospheric Administration. 2019. "Coral Reef Ecosystems." National Oceanic and Atmospheric Administration. https://www.noaa.gov/education/resource-collections/marine-life/coral-reef-ecosystems.

23. National Oceanic and Atmospheric Administration. 2021. "What Is Coral Bleaching?" National Ocean Service. https://oceanservice.noaa.gov/facts/coral_bleach.html.

24. Knowlton, N. 2018. "Corals and Coral Reefs." Smithsonian Ocean. https://ocean.si.edu/ocean-life/invertebrates/corals-and-coral-reefs.

25. Gates, R. 2017. In-person interview.

26. US Fish and Wildlife Service. 2017. "More about Midway." https://www.fws.gov/refuge/Midway_Atoll/more_about_midway.html.

27. US Fish and Wildlife Service. 2017. "More about Midway." https://www.fws
.gov/refuge/Midway_Atoll/more_about_midway.html.
28. Fossi, M. C., et al. 2014. "Large Filter Feeding Marine Organisms as Indicators
of Microplastic in the Pelagic Environment: The Case Studies of the Mediterranean Basking Shark (Cetorhinus Maximus) and Fin Whale (Balaenoptera
Physalus)." *Marine Environmental Research* 100 (September): 17–24. https://
doi.org/10.1016/j.marenvres.2014.02.002.
29. Ovide, B. G. 2019. In-person interview.
30. Fossi, M. C., et al. 2017. "Plastic Debris Occurrence, Convergence Areas and
Fin Whales Feeding Ground in the Mediterranean Marine Protected Area
Pelagos Sanctuary: A Modeling Approach." *Frontiers in Marine Science* 4: 167.
https://doi.org/10.3389/fmars.2017.00167.

## Chapter 4

1. Coniff, R. 2017. "Up from the Depths." *National Wildlife,* December 2017–
January 2018. https://www.nwf.org/Magazines/National-Wildlife/2018/Dec
-Jan/Animals/Vertical-Migration.
2. Møhl, M. 2016. In-person interview.
3. Syberg, K. 2016. In-person interview.
4. Encyclopaedia Britannica. 2020. "Mariana Trench." https://www.britannica
.com/place/Mariana-Trench.
5. Peng, X., et al. 2018. "Microplastics Contaminate the Deepest Part of the
World's Ocean." *Geochemical Perspectives Letters* 9: 1–5. https://www.geo
chemicalperspectivesletters.org/documents/GPL1829_noSI.pdf.
6. Chiba, S., et al. 2018. "Human Footprint in the Abyss: 30 Year Records of
Deep-Sea Plastic Debris." *Marine Policy* 96 (October): 204–12. https://doi.org
/10.1016/j.marpol.2018.03.022.
7. Koelmans, A. A., et al. 2017. "All Is Not Lost: Deriving a Top-down Mass Budget of Plastic at Sea." *Environmental Research Letters* 12 (11): 114028. https://
doi.org/10.1088/1748-9326/aa9500.
8. Barrett, J., et al. 2020. "Microplastic Pollution in Deep-Sea Sediments from the
Great Australian Bight." *Frontiers in Marine Science* 7: 576170. https://doi.org
/10.3389/fmars.2020.576170.
9. Koelmans et al. "All Is Not Lost"; Syberg, K. In-person interview.
10. Syberg, K. In-person interview.

## Chapter 5

1. Encyclopaedia Britannica. 2018. "Marquesas Islands." https://www.britannica
.com/place/Marquesas-Islands.
2. Syberg, K. 2017. In-person interview.
3. Wright, S. L., et al. 2013. "The Physical Impacts of Microplastics on Marine

Organisms: A Review." *Environmental Pollution* 178 (July): 483–92. https://doi
.org/10.1016/j.envpol.2013.02.031.

4. Hartmann, N. B., et al. 2017. "Microplastics as Vectors for Environmental
Contaminants: Exploring Sorption, Desorption and Transfer to Biota." *Inte-
grated Environmental Assessment and Management* 13 (3): 488–93. https://doi
.org/10.1002/ieam.1904.

5. Oehlmann, J., et al. 2009. "A Critical Analysis of the Biological Impacts of Plasticizers on Wildlife." *Philosophical Transactions of the Royal Society B* 364 (1526):
2047–62. https://doi.org/10.1098/rstb.2008.0242.

6. Andrady. A. L. 2011. "Microplastics in the Marine Environment." *Marine
Pollution Bulletin* 62 (8): 1596–1605. https://doi.org/10.1016/j.marpolbul
.2011.05.030.

7. Syberg, K., et al. 2015. "Microplastics: Addressing Ecological Risk through
Lessons Learned." *Environmental Toxicology and Chemistry* 34 (5): 945–53.
https://doi.org/10.1002/etc.2914.

8. Syberg, K. In-person interview.

9. Rochman, C. M., et al. 2013. "Ingested Plastic Transfers Hazardous Chemicals
to Fish and Induces Hepatic Stress." *Scientific Reports* 3 (1). https://doi.org
/10.1038/srep03263.

10. Dale, W. E., et al. 1962. "Storage and Excretion of DDT in Starved Rats." *Toxicology and Applied Pharmacology* 4 (1): 89–106. https://doi.org/10.1016/0041
-008x(62)90078-9.

11. Rochman, C. 2020. Phone interview.

12. Mattson, K., et al. 2017. "Brain Damage and Behavioural Disorders in Fish
Induced by Plastic Nanoparticles Delivered through the Food Chain." *Scientific
Reports* 7 (1). https://doi.org/10.1038/s41598-017-10813-0.

13. Pitt, J. A., et al. 2018. "Uptake, Tissue Distribution and Toxicity of Polystyrene
Nanoparticles in Developing Zebrafish." *Aquatic Toxicology* 194 (January):
185–94. https://doi.org/10.1016/j.aquatox.2017.11.017.

## Chapter 6

1. Mason, S. A. 2017. In-person interview.

2. Rochman, C. M., et al. 2015. "Scientific Evidence Supports a Ban on Microbeads." *Environmental Science and Technology* 49 (18): 10759–61. https://pubs
.acs.org/doi/10.1021/acs.est.5b03909.

3. Rochman, C. M. 2020. Phone interview.

4. Mason, S. A. In-person interview.

5. Napper, I. E., and R. C. Thompson. 2016. "Release of Synthetic Microplastic
Plastic Fibres from Domestic Washing Machines: Effects of Fabric Type and
Washing Conditions." *Marine Pollution Bulletin* 112 (1–2): 39–45. https://
www.sciencedirect.com/science/article/abs/pii/S0025326X16307639.

6. McFall-Johnson, M. 2020. "These Facts Show How Unsustainable the Fashion Industry Is." World Economic Forum. https://www.weforum.org/agenda /2020/01/fashion-industry-carbon-unsustainable-environment-pollution/.

7. Prentice, C. 2014. "Synthetic Fibers Surpass Cotton to Dominate U.S. Apparel Imports." Reuters. https://www.reuters.com/article/usa-cotton-apparel/synthetic -fibers-surpass-cotton-to-dominate-u-s-apparel-imports-idUSL1N0SV2E720 141105.

8. US EPA. 2018. "Textiles: Material-Specific Data." https://www.epa.gov/facts -and-figures-about-materials-waste-and-recycling/textiles-material-specific-data.

9. Lenaker, P. L., et al. 2021. "Spatial Distribution of Microplastics in Surficial Benthic Sediment of Lake Michigan and Lake Erie." *Environmental Science and Technology* 55 (1): 373–84. https://doi.org/10.1021/acs.est.0c06087.

10. Environmental Protection Agency. n.d. "History of the Clean Water Act." https://www.epa.gov/laws-regulations/history-clean-water-act.

11. Alliance for the Great Lakes. 2010. "Protecting the Great Lakes from Pharma-ceutical Pollution." https://cdn.ymaws.com/www.productstewardship.us /resource/collection/FFDF28A1-9926-46E7-87F9-70C7BBD95491/Protecting _the_Great_Lakes_from_Pharmaceutical_Pollution.pdf.

12. US EPA. 2015. "Lake Erie." https://www.epa.gov/greatlakes/lake-erie.

13. NOAA. 2019. "Great Lakes Region Ecosystem-Based Management Activities." https://ecosystems.noaa.gov/WhereIsEBMBeingUsed/GreatLakes.aspx.

14. Mason, S. A. 2019. Phone interview.

15. Mason, S. A., et al. 2020. "High Levels of Pelagic Plastic Pollution Within the Surface Waters of Lakes Erie and Ontario." *Journal of Great Lakes Research* 46 (2): 277–88. https://doi.org/10.1016/j.jglr.2019.12.012.

16. Mason, S. A. In-person interview.

17. Miller, E. 2017. "Proposed EPA Budget Cuts 97% from Great Lakes Resto-ration Initiative." Illinois Public Media. https://will.illinois.edu/news/story /proposed-epa-budget-cuts-97-from-great-lakes-restoration-initiative.

18. Wigdahl-Perry, C. 2020. Phone interview.

19. Madsen, T., et al. 2001. *Environmental and Health Assessment of Substances in Household Detergents and Cosmetic Detergent Products.* Environment project no. 615. Hørsholm, Denmark: Danish Environmental Protection Agency. https:// www2.mst.dk/udgiv/Publications/2001/87-7944-596-9/pdf/87-7944-597-7.pdf.

20. Barrett, H. 2020. Phone interview.

21. Centers for Disease Control and Prevention. 2017. "Per- and Polyfluorinated Substances (PFAS) Factsheet." https://www.cdc.gov/biomonitoring/PFAS _FactSheet.html; Domingo, J. L., and M. Nadal. 2019. "Human Exposure to Per- and Polyfluoroalkyl Substances (PFAS) through Drinking Water: A Review of the Recent Scientific Literature." *Environmental Research* 177 (3): 108648. https://www.sciencedirect.com/science/article/abs/pii/S0013935119304451; Muir, D., et al. 2019. "Levels and Trends of Poly- and Perfluoroalkyl Substances

in the Arctic Environment—An Update." *Emerging Contaminants* 5: 240–71; Strategic Approach to International Chemicals Management. 2019. "Open-Ended Working Group of the International Conference on Chemicals Management: Third Meeting." Montevideo. http://www.saicm.org/Portals/12 /Documents/meetings/OEWG3/inf/OEWG3-INF-28-KMI.pdf.

22. US Environmental Protection Agency. n.d. "What Is Superfund?" https://www .epa.gov/superfund/what-superfund.

23. Mason, S. A. In-person interview.

24. ATSDR. 2021. "Toxicological Profile for Perfluoroalkyls." https://www.atsdr .cdc.gov/toxprofiles/tp200.pdf

25. Mason, S. A. In-person interview.

26. European Environmental Agency. 2019. "Emerging Chemical Risks in Europe —'PFAS.'" https://www.eea.europa.eu/publications/emerging-chemical-risks-in -europe/emerging-chemical-risks-in-europe.

27. Mason, S. A. In-person interview.

28. US EPA. n.d. "Basic Information on PFAS." https://www.epa.gov/pfas/basic -information-pfas.

29. Denchak, M. 2018. "Flint Water Crisis: Everything You Need to Know." https://www.nrdc.org/stories/flint-water-crisis-everything-you-need-know.

30. Sharma, S., and A. Bhattacharya. 2017. "Drinking Water Contamination and Treatment Techniques." *Applied Water Sciences* 7 (3): 1043–67. https://doi.org /10.1007/s13201-016-0455-7.

31. Encyclopaedia Britannica. 2020. "Great Lakes." https://www.britannica.com/ place/Great-Lakes.

32. US EPA. 2019. "Facts and Figures about the Great Lakes." https://www.epa.gov /greatlakes/facts-and-figures-about-great-lakes; Murray, W., et al. 2019. "The Science and Policy of PFAS in the Great Lakes Region: A Roadmap for Local, State and Federal Action." National Wildlife Federation. https://www.nwf.org /-/media/Documents/PDFs/NWF-Reports/2019/NWF-PFAS-Great-Lakes -Region.ashx.

33. Wigdahl-Perry, C. Phone interview.

34. Brush, M. 2013. "'Lake Erie Has 2% of the Water in the Great Lakes, but 50% of the Fish.'" Michigan Public Radio. https://www.michiganradio.org/post/lake -erie-has-2-water-great-lakes-50-fish

35. National Oceanic and Atmospheric Administration. n.d. "Lake Erie Food Web." https://www.glerl.noaa.gov/pubs/brochures/foodweb/LEfoodweb.pdf.

36. Wigdahl-Perry, C. Phone interview.

37. Wigdahl-Perry, C. Phone interview.

38. Lake Erie Foundation. "Invasive Species." https://lakeeriefoundation.org/issues /invasive-species/.

39. Lake George Association. n.d. "Spiny Water Flea." https://www.lakegeorge association.org/educate/science/lake-george-invasive-species/spiny-water-flea/.

## Chapter 7

1. Vianello, A. 2019. In-person interview.

2. Vianello, A. In-person interview.

3. Vianello, A., et al. 2019. "Simulating Human Exposure to Indoor Airborne Microplastics Using a Breathing Thermal Manikin." *Scientific Reports* 9 (1): 8670. https://doi.org/10.1038/s41598-019-45054-w.

4. Bergmann, M. et al. 2019. "White and Wonderful? Microplastics Prevail in Snow from the Alps to the Arctic." *Science Advances* 5 (8): eaax1157. https://advances.sciencemag.org/content/5/8/eaax1157.

5. Wright, S. L., et al. 2019. "Atmospheric Microplastic Deposition in an Urban Environment and an Evaluation of Transport." *Environment International.* https://www.sciencedirect.com/science/article/pii/S0160412019330351.

6. Halle, L. L. 2020. In-person interview.

7. Ryan, J. 2020. "Scientists Point to Chemical in Car Tires That's Been Killing Coho Salmon." Oregon Public Broadcasting. https://www.opb.org/article/2020/12/04/scientists-point-to-chemical-in-car-tires-thats-been-killing-coho-salmon.

8. Vollertsen, J. 2019. In-person interview.

9. Prata, J. C. 2018. "Airborne Microplastics: Consequences to Human Health?" *Environmental Pollution* 234 (March): 115–26. https://doi.org/10.1016/j.envpol.2017.11.043.

10. Oliveri, G., et al. 2020. "Micro- and Nano-Plastics in Edible Fruit and Vegetables. The First Diet Risks Assessment for the General Population." *Environmental Research* 187 (August): 109677. https://doi.org/10.1016/j.envres.2020.109677.

11. Steinmetz, Z., et al. 2016. "Plastic Mulching in Agriculture. Trading Short-Term Agronomic Benefits for Long-Term Soil Degradation?" *Science of the Total Environment* 550 (April): 690–705. https://doi.org/10.1016/j.scitotenv.2016.01.153.

12. de Souza Machado, A. A., et al. 2018. "Impacts of Microplastics on the Soil Biophysical Environment." *Environmental Science Technology.* https://pubs.acs.org/doi/10.1021/acs.est.8b02212.

13. Sun, X. D., et al. 2020. "Differentially Charged Nanoplastics Demonstrate Distinct Accumulation in Arabidopsis Thaliana." *Nature Nanotechnology* 15 (9): 755–60. https://doi.org/10.1038/s41565-020-0707-4.

14. Cox, K. D., et al. 2019. "Human Consumption of Microplastics." *Environmental Science and Technology* 53 (12): 7068–74. https://doi.org/10.1021/acs.est.9b01517.

15. Schwabl, P., et al. 2018. "Assessment of Microplastic Concentrations in Human Stool—Preliminary Results of a Prospective Study." Poster. United European

Gastroenterology Week. https://ueg.eu/library/assessment-of-microplastic
-concentrations-in-human-stool-preliminary-results-of-a-prospective-study
/180360.

16. Syberg, K., et al. 2018. "Health Risk Associated with Plastic Debris on the
Island of Zanzibar: Importance of Associated Pathogenic Bacteria and Implica-
tions for Local Communities." Abstract from Transforming for Sustainability,
Copenhagen, Denmark. https://rucforsk.ruc.dk/ws/portalfiles/portal/64023344
/SYBERG_et_al._Health_risk_associated_with_plastic_debris_on_the_Island
_of_Zanzibar.pdf.

17. Bour, A., et al. 2020. "Microplastic Vector Effects: Are Fish at Risk When
Exposed via the Trophic Chain?" *Frontiers in Environmental Science* 8 (June).
https://doi.org/10.3389/fenvs.2020.00090; Jung, J. W., et al. 2020. "Chronic
Toxicity of Endocrine Disrupting Chemicals Used in Plastic Products in Korean
Resident Species: Implications for Aquatic Ecological Risk Assessment." *Ecotoxi-
cology and Environmental Safety* 192 (April): 110309. https://doi.org/10.1016
/j.ecoenv.2020.110309; European Chemicals Agency. 2011. "Chemicals in
Plastic Products—ECHA." https://chemicalsinourlife.echa.europa.eu/chemicals
-in-plastic-products.

18. Namrata, K., et al. 2020. "Preliminary Results from Detection of Microplastics
in Liquid Samples Using Flow Cytometry." *Frontiers in Marine Science* 7. https://
www.frontiersin.org/articles/10.3389/fmars.2020.552688/full.

19. Carrington, D. 2020. "Microplastic Particles Now Discoverable in Human
Organs." *Guardian.* https://www.theguardian.com/environment/2020/aug/17
/microplastic-particles-discovered-in-human-organs.

20. Ragusa, A., et al. 2021. "Plasticenta: First Evidence of Microplastics in Human
Placenta." *Environment International* 146 (January): 106274. https://doi.org
/10.1016/j.envint.2020.106274.

21. "Trovate per La Prima Volta Microplastiche Nella Placenta Umana." 2020.
*La Repubblica.* https://www.repubblica.it/salute/2020/12/09/news/trovate_per
_la_prima_volta_microplastiche_nella_placenta_umana-277658153.

22. Wright, S. L., et al. 2017. "Plastic and Human Health: A Micro Issue?" *Environmental Science & Technology* 51 (12): 6634–47. https://doi.org/10.1021/acs
.est.7b00423.

23. Niranjan, R., and A. K. Thakur. 2017. "The Toxicological Mechanisms of Environmental Soot (Black Carbon) and Carbon Black: Focus on Oxidative Stress
and Inflammatory Pathways." *Frontiers in Immunology* 8 (June). https://doi.org
/10.3389/fimmu.2017.00763.

24. Hanke, U. M., et al. 2019. "Leveraging Lessons Learned from Black Carbon
Research to Study Plastics in the Environment." *Environmental Science Technology* 53 (12). https://pubs.acs.org/doi/10.1021/acs.est.9b02961/.

## Chapter 8

1. American Lung Association. 2020. "Disparities in the Impact of Air Pollution." https://www.lung.org/clean-air/outdoors/who-is-at-risk/disparities.
2. US Environmental Protection Agency. n.d. "Evolution of the Clean Air Act." https://www.epa.gov/clean-air-act-overview/evolution-clean-air-act.
3. Younes, L. 2019. "What Could Happen If a $9.4 Billion Chemical Plant Comes to 'Cancer Alley.'" ProPublica. https://www.propublica.org/article /what-could-happen-if-a-9.4-billion-chemical-plant-comes-to-cancer-alley; Environmental Protection Agency. 2021. "Greenhouse Gas Equivalencies Calculator." Energy and the Environment. https://www.epa.gov/energy/green house-gas-equivalencies-calculator.
4. ProPublica. 2020. "New Climate Maps Show a Transformed United States." https://projects.propublica.org/climate-migration/.
5. Energy Information Administration. 2020. "Refinery Capacity." http://www.eia .gov/petroleum/refinerycapacity/table5.pdf.
6. Environmental Protection Agency. 2014. "2014 NATA Map." National Air Toxics Assessment. https://www.epa.gov/national-air-toxics-assessment/2014 -nata-map.
7. Lerner, S. *Diamond: A Struggle for Environmental Justice in Louisiana's Chemical Corridor.* Cambridge: MIT Press, 2004, 11–17, 23–28.
8. Nijhuis, M. 2003. "How the Five-Gallon Plastic Bucket Came to the Aid of Grassroots Environmentalists." *Grist.* https://grist.org/article/the19/.
9. Global Nonviolent Action Database. 2002. "Black Residents of Diamond Win Fight with Shell Chemical for Relocation 1989–2002." https://nvdatabase .swarthmore.edu/content/black-residents-diamond-win-fight-shell-chemical -relocation-1989-2002.
10. Wesdock, J. C., and I. M. Arnold. 2014. "Occupational and Environmental Health in the Aluminum Industry: Key Points for Health Practitioners." *Journal of Occupational and Environmental Medicine* 56 (5 Suppl): S5–S11. https://www .ncbi.nlm.nih.gov/pmc/articles/PMC4131940/.
11. "Point Comfort, Texas." United States Census Bureau. https://data.census.gov /cedsci/all?q=point%20comfort%20texss.
12. Chen, C. F., et al. 2018. "Increased Cancer Incidence of Changhua Residents Living in Taisi Village North to the No. 6 Naphtha Cracking Complex." *Journal of the Formosan Medical Association* 117: 12. https://doi.org/10.1016 /j.jfma.2017.12.013.
13. Yuan, T. H., et al. 2018. "Increased Cancers among Residents Living in the Neighborhood of a Petrochemical Complex: A 12-Year Retrospective Cohort Study." *International Journal of Hygiene and Environmental Health* 221 (2): 308–14. https://doi.org/10.1016/j.ijheh.2017.12.004.

14. Martínez, A. L. 2017. "'We Are Jobless Because of Fish Poisoning': Vietnamese Fishermen Battle for Justice." *Guardian*. https://www.theguardian.com/global-development/2017/aug/14/vietnamese-fishermen-jobless-fish-poisoning-battle-justice; Fernández, S. 2019. "Plastic Company Set to Pay $50 Million Settlement in Water Pollution Suit Brought on by Texas Residents." *Texas Tribune*. https://www.texastribune.org/2019/10/15/formosa-plastics-pay-50-million-texas-clean-water-act-lawsuit/; Vass, K. 2020. "From Louisiana to Taiwan, Environmental Activists Stand up to a Major Plastics Company." *The World from PRX*. https://www.pri.org/stories/2020-07-09/louisiana-taiwan-environmental-activists-stand-major-plastics-company.

15. Edwards, J. B. 2020. "RE: Veto of House Bill 197 of the 2020 Regular Session." https://gov.louisiana.gov/assets/docs/Vetos-Regular/Veto-HB-197.pdf.

## Chapter 9

1. Ripple, William J., et al. 2020. "World Scientists' Warning of a Climate Emergency." *BioScience* 70 (1): 8–12. https://doi.org/10.1093/biosci/biz088; Tong, D., Q. Zhang, Y. Zheng, et al. 2019. "Committed Emissions from Existing Energy Infrastructure Jeopardize 1.5 °C Climate Target." *Nature* 572: 373–77. https://doi.org/10.1038/s41586-019-1364-3.

2. Shepard, P. 2021. Email correspondence.

3. Le Quéré, C., et al. 2020. "Temporary Reduction in Daily Global CO2 Emissions During the COVID-19 Forced Confinement." *Nature Climate Change* 10: 647–35. https://www.nature.com/articles/s41558-020-0797-x.

4. United Nations. 2019. "Emissions Gap Report 2019." https://www.unep.org/resources/emissions-gap-report-2019.

5. US Bureau of Labor Statistics. 2020. "From the Barrel to the Pump: The Impact of the COVID-19 Pandemic on Prices for Petroleum Products." Monthly Labor Review. https://www.bls.gov/opub/mlr/2020/article/from-the-barrel-to-the-pump.htm.

6. Gardiner, B. 2020. "Pollution Made COVID-19 Worse. Now Lockdowns Are Cleaning the Air." *National Geographic*. https://www.nationalgeographic.com/science/article/pollution-made-the-pandemic-worse-but-lockdowns-clean-the-sky.

7. World Meteorological Organization. 2020. "United in Science 2020." https://public.wmo.int/en/resources/united_in_science.

8. McKibben, B. 2020. "Are We Past the Peak of Big Oil's Power?" *New Yorker*. https://www.newyorker.com/news/annals-of-a-warming-planet/are-we-past-the-peak-of-big-oils-power.

9. American Chemistry Council. 2018. "US Chemical Industry Investment Linked to Shale Gas Reaches $200 Billion."

10. World Economic Forum. 2016. "The New Plastics Economy: Rethinking the Future of Plastics." http://www3.weforum.org/docs/WEF_The_New_Plastics _Economy.pdf.

11. Associated Press. 2008. "Y. C. Wang, Billionaire Who Led Formosa Plastics, Is Dead at 91." https://www.nytimes.com/2008/10/17/business/17wang.html.

12. Tullo, A. H. 2020. "C&EN's Global Top 50 for 2020." *American Chemical Society.* https://www.cen.acs.org/business/finance/CENs-Global-Top-50-2020 /98/i29.

13. International Energy Agency. 2018. "The Future of Petrochemicals." https:// www.iea.org/reports/the-future-of-petrochemicals.

14. Center for International Environmental Law. 2019. "Plastic and Climate: The Hidden Costs of a Plastic Planet." https://www.ciel.org/plasticandclimate/.

15. Royer, S. J., et al. 2018. "Production of Methane and Ethylene from Plastic in the Environment." *PLOS ONE* 13 (8). https://doi.org/10.1371/journal .pone.0200574.

16. Kelly, A., et al. 2020. "Microplastic Contamination in East Antarctica Sea Ice. *Marine Plastic Bulletin* 154. https://doi.org/10.1016/j.marpolbul.2020.111130; Haram, L. E., et al. 2020. "A Plastic Lexicon." *Marine Pollution Bulletin* 150. https://doi.org/10.1016/j.marpolbul.2019.110714.

17. Soper, F. M., et al. "Leaf-Cutter Ants Engineer Large Nitrous Oxide Hot Spots in Tropical Forests." *Proceedings of the Royal Society B* 286. https://doi.org /10.1098/rspb.2018.2504.

18. Altman, R. 2018. "American Beauties." *Topic.* https://www.topic.com/american -beauties.

19. Altman, R. 2019. Phone interview.

20. Rothstein, S. I. 1973. "Plastic Particle Pollution of the Surface of the Atlantic Ocean: Evidence from a Seabird." *The Condor* 75 (3). https://doi.org/10.2307 /1366176.

21. Scott, G. 1972. "Plastics Packaging and Coastal Pollution." *International Journal of Environmental Studies.* https://doi.org/10.1080/00207237208709489.

22. Carpenter, E. J. 2019. Email correspondence; Carpenter, E. J., and K. L. Smith. 1972. "Plastics on the Sargasso Sea Surface." *Science* 175 (4027): 1240–41. https://science.sciencemag.org/content/175/4027/1240; Carpenter, E. J., et al. 1972. "Polystyrene Spherules in Coastal Waters." *Science* 178 (4062): 749–50. https://science.sciencemag.org/content/178/4062/749.

23. Google Scholar. 2021. https://scholar.google.com/scholar?q=microplastic&hl =en&as_sdt=0%2C33&as_ylo=2020&as_yhi=2020.

24. Knapman, C., et al. 2000. "Proceedings of the International Marine Debris Conference: Derelict Fishing Gear in the Ocean Environment." National Oceanic and Atmospheric Administration. https://nmshawaiihumpbackwhale.blob .core.windows.net/hawaiihumpbackwhale-prod/media/archive/documents/pdfs _conferences/proceedings.pdf.

25. Knapman et al. "Proceedings of the International Marine Debris Conference."
26. Sullivan, L. 2020. "How big oil misled the public into believing plastic would be recycled." NPR. https://wamu.org/story/20/09/11/how-big-oil-misled-the -public-into-believing-plastic-would-be-recycled/.
27. Shiver, J. 1986. "Supermarket Dilemma: Battle of the Bags: Paper or Plastic?" *Los Angeles Times.* https://www.latimes.com/archives/la-xpm-1986-06-13-mn -10728-story.html.
28. Belkin, L. 1984. "Battle of the Grocery Bags: Plastic vs. Paper." *New York Times.* https://www.nytimes.com/1984/11/17/style/battle-of-the-grocery-bags-plastic -versus-paper.html.
29. Altman, R. 2019. Phone interview.
30. Kelechava, B. 2019. "Resin Identification Codes (RICs), As Specified by ASTM D7611." https://blog.ansi.org/2019/02/resin-identification-codes-rics-astm -d7611/#gref.
31. Law, K. L., et al. 2020. "The United States' Contribution of Plastic Waste to Land and Ocean." *Science Advances* 6 (44): eabd0288. https://doi.org/10.1126 /sciadv.abd0288.
32. McCormick, E., et. al. 2019. "Where Does Your Plastic Go? Global Investiga- tion Reveals America's Dirty Secret." *Guardian.* https://www.theguardian.com /us-news/2019/jun/17/recycled-plastic-america-global-crisis.
33. Verma, R., et al. 2016. "Toxic Pollutants from Plastic Waste—A Review." *Proce- dia Environmental Sciences* 35: 701–8. https://www.sciencedirect.com/science /article/pii/S187802961630158X.
34. Wilson, D. C., et al. "Role of Informal Sector Recycling in Waste Management in Developing Countries." *Habitat International* 4 (30). https://www.science direct.com/science/article/pii/S0197397505000482.
35. Brooks, A. L., et al. 2018. "The Chinese Import Ban and Its Impact on Global Plastic Waste Trade." *Scientific Advances* 4 (6). https://advances.sciencemag.org /content/4/6/eaat0131.
36. Plastics Europe. 2018. "Plastics—The Facts." https://www.plasticseurope.org /application/files/6315/4510/9658/Plastics_the_facts_2018_AF_web.pdf.
37. UN Comtrade Database. http://www.diken.com.tr/15-yilda-173-kat-artti-her -gun-213-kamyon-plastik-atik-ulkemize-bosaltiliyor/.
38. Union of Concerned Scientists. 2017. "The Disinformation Playbook: How Business Interests Deceive, Misinform, and Buy Influence at the Expense of Public Health and Safety." https://www.ucsusa.org/resources/disinformation -playbook.
39. Gunn, K. 2018. "Danes Use Far Fewer Plastic Bags Than Americans—Here's How." *National Geographic.*
40. Law, K. L., et. al. 2020. "The United States' Contribution of Plastic Waste to Land and Ocean." *Science Advances* 6 (44). https://advances.sciencemag.org /content/6/44/eabd0288.

41. Eriksen, M. 2019. Email correspondence.
42. Moore, C. 2017. In-person interview.
43. Altman, R. 2019. Phone interview.
44. Carpenter, E. J. 2019. Email correspondence; Moore, C. 2009. "Seas of Plastic." TED video. https://www.ted.com/talks/charles_moore_seas_of_plastic; Jordan, C. 2018. *Albatross*; Figgener, C. 2015. "Sea Turtle with Straw Up Its Nostril— 'No' to Plastic Straws." YouTube. https://www.youtube.com/watch?v=4wH878t 78bw.
45. Carpenter, E. J., and K. L. Smith. 1972. "Plastics on the Sargasso Sea Surface." *Science* 175 (4027).

## Chapter 10

1. Borrelle, S. B., et al. 2020. "Predicted Growth in Plastic Waste Exceeds Efforts to Mitigate Plastic Pollution." *Science* 369 (6510): 1515–18. https://www.science.sciencemay.org/content/369/6510/1515.
2. Ocean Conservancy. 2019. "Ahead of the 34th International Coastal Cleanup, Ocean Conservancy Report Reveals Prevalence of Plastic Cutlery on Beaches, Waterways." https://oceanconservancy.org/news/ahead-34th-international -coastal-cleanup-ocean-conservancy-report-reveals-prevalence-plastic-cutlery -beaches-waterways/.
3. Stoksad, E. 2018. "Controversial Plastic Trash Collector Begins Maiden Ocean Voyage." *Science*. https://www.sciencemag.org/news/2018/09/still-controversial -plastic-trash-collector-ocean-begins-maiden-voyage.
4. The Ocean Cleanup. 2021. "FAQ." https://theoceancleanup.com/faq/.
5. The Seabin Project. 2021. "Seabin Project's 'Whole Solution' Proposal for Ocean Conservation and Sustainability." https://seabinproject.com/seabin -projects-whole-solution-proposal-for-ocean-conservation-and-sustainability/.
6. Burris, J. 2015. "Water Wheel Scoops 19 Tons of Inner Harbor Trash in One Day." *Baltimore Sun*. https://www.baltimoresun.com/news/environment/bs-md -ci-water-wheel-trash-0423-20150422-story.html.
7. University of Queensland, Australia. 2015. "World's Turtles Face Plastic Deluge Danger." UQ News. https://www.uq.edu.au/news/article/2015/09/ world%E2%80%99s-turtles-face-plastic-deluge-danger.
8. Wilcox, C., et al. 2018. "A Quantitative Analysis Linking Sea Turtle Mortality and Plastic Debris Ingestion." *Scientific Reports* 8 (12536). https://www.doi .org/10.1038/541598-018-30038-z.
9. Jensen, M. P., et al. 2018. "Environmental Warming and Feminization of One of the Largest Sea Turtle Populations in the World." *Current Biology* 28 (1). https://pubmed.ncbi.nlm.nih.gov/29316410.; Duncan, Emily M., et. al. 2018. "The True Depth of the Mediterranean Plastic Problem: Extreme Microplastic Pollution on Marine Turtle Nesting Beaches in Cyprus." *Marine Pollution Bulletin* 136. https://doi.org/10.1016/j.marpolbul.2018.09.019.

10. Lee, P. J., and J. K. Willis. 1999. *Ho'opono*. Honolulu: Night Rainbow Publishing.
11. Beckwith, M. W. 1970. *Hawaiian Mythology*. Honolulu: University of Hawai'i Press.
12. National Park Service. 2019. "Hawai'i Island: Hawai'i Volcanoes National Park." https://www.nps.gov/havo/index.htm.
13. Güentzel, S. 2016. "Microplastics III/Discofish." http://www.swaantje-guentzel .de/#/microplastics-iii-discofish.
14. *Phuket News*. 2019. "Phuket Most Tourists Per Square Mile in World, Says Report." https://www.thephuketnews.com/phuket-most-tourists-per-square -mile-in-world-says-report-72949.php.
15. Law, K. L., et al. 2020. "The United States' Contribution of Plastic Waste to Land and Ocean." *Science Advances* 6 (44). https://advances.sciencemag.org /content/6/44/eabd0288.

## Chapter 11

1. Tabone, M. D., et al. 2010. "Sustainability Metrics: Life Cycle Assessment and Green Design in Polymers." *Environmental Science and Technology* 44 (21).
2. Greene, J. 2018. Phone interview.
3. Mugica, Y., and A. S. Collins. 2017. "From Fork to Farm and Back: San Francisco Composting." Natural Resources Defense Council. https://www.nrdc.org /resources/san-francisco-composting; US Environmental Protection Agency. 2021. "Zero Waste Case Study: San Francisco." *Managing and Transforming Waste Streams—A Tool for Communities*. https://www.epa.gov/transforming -waste-tool/zero-waste-case-study-san-francisco.
4. Department of Primary Industries and Regional Development. 2018. "Composting to Avoid Methane Production." Government of Western Australia. https://www.agric.wa.gov.au/climate-change/composting-avoid-methane -production.
5. Nace, T. 2019. "Thailand Supermarket Ditches Plastic Packaging for Banana Leaves." *Forbes*. https://www.forbes.com/sites/trevornace/2019/03 /25/thailand-supermarket-uses-banana-leaves-instead-of-plastic-packaging /?sh=41f36a037102.
6. US Environmental Protection Agency. 2019. "Advancing Sustainable Materials Management: 2017 Fact Sheet. (EPA 530-F-19-007)." https://www.epa.gov /sites/production/files/2019-11/documents/2017_facts_and_figures_fact_sheet _final.pdf.
7. US AID. 2020. "Countries with TFCA Programs." https://www.usaid.gov/.
8. Raworth, K. 2017. *Doughnut Economics: Seven Ways to Think Like a 21st-Century Economist*. London: Random House.
9. US Environmental Protection Agency. 2021. "Zero Waste Case Study: Berkeley." *Managing and Transforming Waste Streams—A Tool for Communities*. https://www.epa.gov/transforming-waste-tool/zero-waste-case-study-berkeley;

Pellissier, H. 2010. "Urban Ore Ecopark." *New York Times.* https://www
.nytimes.com/2010/09/26/us/26bcintel.html.

10. Business Wire. 2019. "More Than Half of Consumers Would Pay More for Sustainable Products Designed to Be Reused or Recycled, Accenture Survey Finds." *Berkshire Hathaway.* https://www.businesswire.com/news/home/20190604
005649/en.

11. UN Environment Programme Stockholm Convention. 2021. "Status of Ratification." http://www.pops.int/Countries/StatusofRatifications/Partiesand
Signatoires/tabid/4500/Default.aspx.

12. UN Environment Programme Stockholm Convention. 2021. "All POPs listed
in the Stockholm Convention." http://chm.pops.int/TheConvention/ThePOPs
/AllPOPs/tabid/2509/Default.aspx.

13. BBC News. "UN Resolution Pledges to Plastic Reduction by 2030." BBC.
https://www.bbc.com/news/science-environment-47592111.

14. Bhalla, N., and J. Ndiso. 2019. "U.S. Weakens First Global Commitment on
Curbing Single-Use Plastics." Reuters. https://www.reuters.com/article/us-global
-plastics-pollution/u-s-weakens-first-global-commitment-on-curbing-single-use
-plastics-idUSKCN1QW2J7.

15. McVeigh, K. 2020. "Global Treaty to Tackle Plastic Pollution Gains Steam
without US and UK." *Guardian.* https://www.theguardian.com/environment
/2020/nov/16/us-and-uk-yet-to-show-support-for-global-treaty-to-tackle-plastic
-pollution.

16. Global Legislative Toolkit. 2021. "Legislative Data Sets and Maps." https://
plasticpollutioncoalitionresources.org/resources/maps/.

17. McCarthy, J., and P. Gralki. 2019. "Tourists Banned from Bringing Plastic Bags
to Tanzania." Global Citizen. https://www.globalcitizen.org/en/content/tanzania
-plastic-bag-ban-travelers/.

18. UN Environment Programme. 2019. "Plastic Bag Bans Can Help Reduce Toxic
Fumes." https://www.unep.org/news-and-stories/story/plastic-bag-bans-can
-help-reduce-toxic-fumes.

19. 114th Congress. 2015. "H.R.1321—Microbead-Free Waters Act of 2015."
https://www.congress.gov/bill/114th-congress/house-bill/1321.

20. PlasticBagLaws.Org. 2020. "Preemption Laws." https://www.plasticbaglaws.org
/preemption.

21. House of Representatives, 26th Legislature, State of Hawai'i. 2012. "H.B.
No. 1828: A Bill for an Act." https://www.capitol.hawaii.gov/session2012/bills
/HB1828_.HTM.

22. City Council, City and County of Honolulu, Honolulu, Hawaii. 2016. "Ordinance 17-37, Bill 59 (2016), FD1, CD3." http://www.opala.org/solid_waste
/pdfs/ORD%2017-37%20PBB.pdf.

23. US Environmental Protection Agency. 2015. "Versions of the Waste Reduction

Model (WARM)." https://www.epa.gov/warm/versions-waste-reduction-model -warm.

24. Children's Environmental Health Network. 2018. "FAQs Styrofoam." https:// cehn.org/styrofoam-faq-final-july-2018/; US Environmental Protection Agency. 2000. "Styrene: Hazard Summary." https://www.epa.gov/sites/production/files /2020-05/documents/styrene_update_2a.pdf.

25. World Centric. 2021. "Certifications." https://www.worldcentric.com/our -impact/zero-waste-solutions/certifications/.

26. Zimmermann, L., et. al. 2020. "Are Bioplastics and Plant-Based Materials Safer Than Conventional Plastics? In Vitro Toxicity and Chemical Composition." *Environment International* 145. https://doi.org/10.1016/j.envint.2020.106066.

27. State of New Jersey Governor Phil Murphy. 2020. "Governor Murphy Signs Legislation Banning Single-Use Paper and Plastic Bags in New Jersey." https:// www.nj.gov/governor/news/news/562020/20201104a.shtml.

28. Knott, B. C., et al. 2020. "Characterization and Engineering of a Two-Enzyme System for Plastics Depolymerization." *PNAS* 117 (41). https://doi.org/10.1073 /pnas.2006753117.

29. Drahl, C. "Plastics Recycling with Microbes and Worms Is Further Away Than People Think." *Chemical & Engineering News* 96 (25). https://cen.acs.org /environment/sustainability/Plastics-recycling-microbes-worms-further/96/i25.

## Chapter 12

1. European Commission. 2019. "Directive (EU) 2019/904 of the European Parliament and of the Council of 5 June 2019 on the Reduction of the Impact of Certain Plastic Products on the Environment." https://eur-lex.europa.eu/eli /dir/2019/904/oj.

2. C40 Cities. 2019. "Circular Copenhagen—70% Waste Recycled by 2024." https://www.c40.org/case_studies/circular-copenhagen-70-waste-recycled-by -2024.

3. McKinsey and Company. 2019. "The New Plastics Economy: A Danish Research, Innovation and Business Opportunity." https://www.mckinsey.com /featured-insights/europe/the-new-plastics-economy-a-danish-research -innovation-and-business-opportunity.

4. Murray, A. 2019. "The Incinerator and the Ski Slope Tackling Waste." BBC. https://www.bbc.com/news/business-49877318.

5. Eunomia. 2019. "Nordic Region Must Increase Recycling, Report Finds." https://www.eunomia.co.uk/nordic-region-must-increase-recycling-report-finds/.

6. Network for Circular Packaging. 2019. "Design Guide: Reuse and Recycling of Plastic Packaging." https://plast.dk/wp-content/uploads/2019/12/Design -Guide-Reuse-and-recycling-of-plastic-packaging-for-private-consumers-english -version-1.pdf; OECD. 2019. "OECD Environmental Performance Reviews:

Denmark." https://www.oecd-ilibrary.org/sites/d1eaaba4-en/index.html?itemId =/content/component/d1eaaba4-en.

7. C40 Cities. "Circular Copenhagen."

8. Regional Activity Centre for Sustainable Consumption and Production, et al. 2020. "Plastic's Toxic Additives and the Circular Economy." http://www.cprac .org/en/news-archive/general/toxic-additives-in-plastics-hidden-hazards-linked -to-common-plastic-products.

9. Møhl, M. 2020. Personal communication. In-person interview.

10. McKinsey and Company. "New Plastics Economy."

11. Circular Copenhagen. 2020. "Recycling of 250.000 Plastic Food Trays from Copenhagen." https://circularcph.cphsolutionslab.dk/cc/news/recycling-of-250 -000-plastic-food-trays-from-copenhagen; Jensen, J. E. 2020. LinkedIn Update. https://www.linkedin.com/feed/update/urn:li:activity:6696320859679842304/.

12. British Plastics Federation, et al. 2020. "Recycled Content Used in Plastic Pack-aging." https://cdnmedia.eurofins.com/corporate-eurofins/media/12153132/bpf -recycled-content-us.pdf.

13. Mondelez International. 2019. "Recyclable Packaging." https://www.mondelez international.com/News/100-Recyclable-Packaging; Staub, C. 2020. "Major Packaging Users Hit 6.2% Average Recycled Content." https://resource -recycling.com/plastics/2020/11/11/major-packaging-users-hit-6-2-average -recycled-content/.

14. European Commission. 2018. "Directive of the European Parliament and of the Council on the Reduction of the Impact of Certain Plastic Products on the Environment." https://eur-lex.europa.eu/legal-content/EN/TXT/?uri=CELEX: 52018PC0340; Ries, F. 2019. "European Parliament Votes for Single-Use Plastics Ban." https://ec.europa.eu/environment/efe/news/european-parliament -votes-single-use-plastics-ban-2019-01-18_en.

15. PlasticsEurope. 2019. "Plastics—The Facts 2019." https://www.plasticseurope .org/application/files/1115/7236/4388/FINAL_web_version_Plastics_the _facts2019_14102019.pdf.

16. Coca-Cola Company, et al. 2018. "Single-Use Plastics—Alternative Proposal to Address the Littering of Beverage Caps." Letter to EU Commission. https:// www.euractiv.com/wp-content/uploads/sites/2/2018/10/Letter-on-tethered-cap -alternative-solution_Council.pdf.

17. US Congress, *Break Free from Plastic Pollution Act of 2020*, HR 5845, 116th Cong., introduced in House February 11, 2020, https://www.congress.gov/bill /116th-congress/house-bill/5845; Volcovici, V. 2021. "US Lawmakers Target Plastic Pollution, Producers in New Legislation." https://www.reuters.com/ article/usa-environment-plastic/u-s-lawmakers-target-plastic-pollution-pro ducers-in-new-legislation-idUSL1N2LM2GG.

18. Scientist Action and Advocacy Network. "Effectiveness of Plastic Regulation Around the World." https://scaan.net/plastic_global/.

## Conclusion

1. *Merriam-Webster*. 2021. "plastic (*n.*)." https://www.merriam-webster.com/dictionary/plastic.

2. American Chemical Society. 1993. "National Historic Chemical Landmarks. Bakelite: The World's First Synthetic Plastic." https://www.acs.org/content/acs/en/education/whatischemistry/landmarks/bakelite.html.

3. American Chemistry Council. 2015. "How Can the USA Dramatically Increase Recycling Rates?" *Plastics Engineering*. http://read.nxtbook.com/wiley/plastics engineering/march2015/howcatheusa.html.

4. NASA Orbital Debris Program Office. "Frequently Asked Questions." https://orbitaldebris.jsc.nasa.gov/faq/.

5. Wieser, H. 2016. "Beyond Planned Obsolescence: Product Lifespans and the Challenges to a Circular Economy." *GAIA—Ecological Perspectives for Science and Society* 25 (3): 156–60. https://doi.org/10.14512/gaia.25.3.5.

6. Berry, W. 2010. *What Are People For?* Berkeley, CA: Counterpoint Press.

7. *Center for Biological Diversity, et al. v. US Army Corps of Engineers, et al. and FG LA LLC*. 2020. Case No.: 1:20-cv-00103-RDM. United States District Court for the District of Columbia. https://www.biologicaldiversity.org/programs/climate_law_institute/pdfs/2020-07-14-Prelim-Inj-Brief-Formosa-Plastics.pdf.

8. US Army Corps of Engineers. 2021. "Mission and Vision." https://www.usace.army.mil/About/Mission-and-Vision/.

9. Dermansky, J., and S. Kelly. 2020. "Following Lawsuit, Formosa Agrees to Hold Major Construction on One of Largest Planned US Plastics Plants Until 2021." *DeSmog*. https://www.desmog.com/2020/07/24/formosa-plastics-petro chemical-plants-st-james-delay-construction-lawsuit/.

10. United Nations Human Rights: Office of the High Commissioner. 2021. "USA: Environmental Racism in "Cancer Alley" Must End—Experts." https://www.ohchr.org/EN/NewsEvents/Pages/DisplayNews.aspx?NewsID=26824&LangID=E.

11. Carrington, D. 2021. "Climate Crisis: Record Ocean Heat in 2020 Supercharged Extreme Weather." *Guardian*. https://www.theguardian.com/environment/2021/jan/13/climate-crisis-record-ocean-heat-in-2020-supercharged-extreme-weather; Hassan, F., and E. Peltier. 2020. "Scorching Temperatures Bake Middle East Amid Eid al-Adha Celebrations." *New York Times*. https://www.nytimes.com/2020/07/31/world/middleeast/Middle-East-heat-wave.html; Pike, L. 2020. "What Wildfires in Brazil, Siberia, and the US West Have in Common." *Vox*. https://www.vox.com/21441711/2020-california-wildfires-brazil-amazon-pantanal-siberia-climate-change.

12. Christmas, M. 2017. "How the Warming World Could Turn Many Plants and Animals into Climate Refugees." https://theconversation.com/how-the-warming-world-could-turn-many-plants-and-animals-into-climate-refugees-72722.

13. McDonnell, T. 2018. "The Refugees the World Barely Pays Attention To."

NPR. https://www.npr.org/sections/goatsandsoda/2018/06/20/621782275/the-refugees-that-the-world-barely-pays-attention-to.

14. Global Data Energy. 2019. "North America Has the Highest Oil and Gas Pipeline Length Globally." *Offshore Technology.* https://www.offshore-technology.com/comment/north-america-has-the-highest-oil-and-gas-pipeline-length-globally/; NASA Science. 2021. "Earth: By the Numbers." Solar System Exploration." https://solarsystem.nasa.gov/planets/earth/by-the-numbers/.

## About the Author

Photo by Rasmus Hytting

Erica Cirino is a science writer and artist exploring the intersection of the human and nonhuman worlds. Her widely published photojournalistic and creative works depict the numerous ways people connect to nature—wild creatures in particular—and shape planet Earth. One of her major inspirations is her role as a licensed wildlife rehabilitator who has cared for thousands of sick, orphaned, and injured wild animals in preparation for their eventual release back into their natural habitats. Through her writing, art, and wildlife rehabilitation work, Cirino strives to foster human care, conversation, and kinship with the nonhuman world. She is a recipient of fellowships from Woods Hole Oceanographic Institution, Craig Newmark Graduate School of Journalism at CUNY, and the Safina Center. She lives with her rescued street dog, Sabi, on and between two shores: Long Island and Connecticut.